Lincoln,

the War President

Lincoln,
the War President

The Gettysburg Lectures

edited by GABOR S. BORITT

essays by

ROBERT V. BRUCE

JAMES M. McPHERSON

DAVID BRION DAVIS

CARL N. DEGLER

KENNETH M. STAMPP

ARTHUR M. SCHLESINGER, JR.

GABOR S. BORITT

New York · Oxford · Oxford University Press · 1992

Oxford University Press

Oxford   New York   Toronto
Delhi   Bombay   Calcutta   Madras   Karachi
Kuala Lumpur   Singapore   Hong Kong   Tokyo
Nairobi   Dar es Salaam   Cape Town
Melbourne   Auckland

and associated companies in
Berlin   Ibadan

Copyright © 1992 by Gabor S. Boritt

Published by Oxford University Press, Inc.,
200 Madison Avenue, New York, New York 10016

Library of Congress Cataloging-in-Publication Data
Lincoln, the war president : the Gettysburg lectures / edited by Gabor
S. Boritt ; essays by Robert V. Bruce . . . [et al.].
p. cm.   Includes bibliographical references.
ISBN 0-19-507891-8
1. Lincoln, Abraham, 1809–1865.   2. United States—Politics and
government—Civil War, 1861–1865.   I. Boritt, G. S., 1940–
II. Bruce, Robert V.
E457.L76 1992
973.7′092—dc20   92-19696

The Langston Hughes excerpt (p. vii) is from *The Dream Keeper and
Other Poems* by Langston Hughes. Copyright 1932 by Alfred A. Knopf,
Inc. and renewed 1960 by Langston Hughes. Reprinted by permission of
the publisher.

Frontispiece: Watercolor by Rea Redifer
of Chadds Ford, 1992.

9 8 7 6 5 4 3 2 1

Printed in the United States of America
on acid-free paper

for
ROBERT BRUCE
who taught me history

Acknowledgments

Americans interested in history need to make the pilgrimage to Gettysburg, at least in a symbolic sense, and come to terms with Abraham Lincoln. He is, in the words of Langston Hughes,

> Quiet—
> And yet a voice forever
> Against the
> Timeless walls
> Of time—

For historians the imperatives are the same—but, if anything, stronger. And so every year for the anniversary of Lincoln's address, Gettysburg College invites a looming figure of American scholarship to deliver the Robert M. Fortenbaugh Memorial Lecture, named after a faithful professor who taught there for thirty-six years, principally during the first half of the twentieth century. The charge is to speak to the literate general public without slipping solid scholarly moorings. Any issue of the Civil War era is fair game, but a remarkable number of the historians are seduced by the magic of Gettysburg to look for Lincoln. They tend to consider large, important questions and, as year follows year, develop something of a dialogue. Though as many as a thousand people—some coming from as far as California and Texas—have attended these lectures, and they have been printed privately, this is the first time they are brought before the larger public, with each lecture slightly revised.

The one exception in this book drawn from the last nine years of Fortenbaugh lectures is my own "War Opponent and War President." I am painfully aware of the presumption of putting myself in the company of my betters, including my own teacher Robert V. Bruce. Yet, delivered as the inaugural lecture for the first fully funded chair in the country for the study of the Civil War, the chapter fits the book. It, too, is looking for Lincoln at Gettysburg.

In my acknowledgment of many debts, the historians contributing to this volume must come first. One and all they took their task with great seriousness. Working with them—some of the finest historians in America today—has been a pure pleasure. I hope that they, too, enjoyed the road to Gettysburg.

For thirty years now the Fortenbaugh Lecture has been supported by Gettysburg College. Its commitment to be a guardian of history deserves respect. My congenial colleagues at the History Department and on many other parts of the campus also deserve thanks for their support and friendship.

The able staff at the Civil War Institute helps organize the Fortenbaugh lectures each November 19 and, under the leadership of Tina Fair, helped get this book ready for the publisher. But Charlene Locke, Linda Marshall, earlier Lucie Wolfe, and my research assistant Robert Sandow all had a hand in the book. I also wish to acknowledge the help of the Gettysburg College Civil War Club over the years, especially John Stoudt, Steven Herr, Craig Montesano, Kim Johnson, Ellen Abrahamson, Christina Ericson, Jennifer Haase, Albert Pennino, Patricia Taylor, and Peter Vermilyea.

In the adventurous and torturous task of procuring illustrations, I received help from David Brion Davis of Yale University, Noble Cunningham of the University of Missouri, Columbia, Amanda-Jane Doran of the *Punch* Library, London, Julie Hardwick and David Hedrick of Gettysburg College, James R. Mellon II, Mark E. Neely, Jr. of the Lincoln Library, Thomas F. Schwartz of the Illinois State Historical Library, Michael J. Winey of the U.S. Army Military History Institute, and Linda Ziemer of the Chicago Historical Society. Finally, Rea Redifer of Chadds Ford, whose memorable watercolors provide the covers for the Fortenbaugh Lectures, created the startling war president, a man in pain, for the frontispiece.

Mark Neely helped inspire the private printing of the Gettysburg lectures and made excellent suggestions for editing Chapter 6. My editors at Oxford, Sheldon Meyer and Leona Capeless, again proved to be a pleasure to work with.

My family, Liz, Norse, Jake, and Daniel, are indispensable to all my work. Norse also created, once again, a dedication page using Pennsylvania Dutch motifs, with some other delights smuggled in as well. Jake, in addition to working on this volume and at the Civil War Institute, created for the jacket the imaginative photograph of his father with Lincoln.

The book is dedicated to Bob Bruce, once and always my teacher, and my oldest friend in the historical profession.

Each chapter in *Lincoln, the War President* stands on its own (in spite of cross references) and the reader can safely jump from Bruce to Schlesinger and back to McPherson. However, together the chapters make a book that sheds intriguing new light on both Lincoln and the Civil War, a book that will also be provocative to many readers. There is danger in making such a unifying claim about a work of seven strong-minded historians, even if the variety in their thought and styles (down to the footnotes, standardized forms notwithstanding) is one of the book's pleasures. It reminds me of Lincoln's comment about the Yankee peddler who sold pantaloons "big enough for any man, small enough for any boy." But there is only one good way for a reader to decide about the book— like Lincoln's pantaloons—try it on and see if it fits.

Farm by the Ford   G. S. B.
Gettysburg
Spring 1992

Contents

Illustrations

Introduction

Lincoln walked up to the parapet in full view of the Confederates a hundred yards away. Bullets flew around him but he stood there, six foot four, top hat on his head to appear taller. All knew on both sides who the man was. The bullets whistled. An officer close to him went down. But Lincoln just stood there. Looking. In the July heat of 1864, in the fourth year of the war, Washington was under attack.

Was it ever going to end, this summer of discontent with its soon to be 100,000 casualties, this war with its million and a half? Before the summer was over the president would pen a memorandum to himself saying that he did not expect to be reelected.[1] The future of the United States, black people, Lincoln—all looked awesomely bleak.

And what was he doing at the parapet of Fort Stevens while the forces of General Jubal Early attacked the capital? Chapter

Seven ventures an answer: Lincoln stood there "looking not at the Confederates, but God, saying silently: if I am wrong, God, strike me down." Ordered down roughly by an officer, and called a fool, Lincoln was later found sitting with his back against the parapet. Looking.

This episode is a slight matter of little consequence. After all Lincoln lived—at least into the next spring. Yet the tale makes the point, in a small way, that this is an often audacious book, a quality Lincoln had and admired, one that he as a subject deserves from historians. The seven scholars in this book will not disappoint readers on this score.

We consider large questions, often of global meaning and consequence. We ask hard questions. Even as we look for Lincoln we move readily from Latin America to Japan, to Germany, to Africa, and elsewhere. In terms of time we move from antiquity to our very own days.

In Chapter One, "The Shadow of a Coming War," Robert Bruce starts with the era of the American founding. He could, like other historians, lament that we look too much into the unknown future, and learn all too little from the known past. Instead he takes his premise from the common sense notion that in history, as in everyday life, we act both on "our perceptions of the past" and "our expectations of the future." Bruce then assesses how the future appeared to Americans during the eighty-some years between the uniting of the nation in the later 1700s and its coming apart in 1861. Specifically, he focuses on American premonitions of civil war.

Bruce takes the familiar, spices it with the little known and the new, and weaves a fresh pattern that shows the long and frightening shadow of a coming war. We hear the voices of those years—as Americans of yore heard the rumors of war— voices of fools and voices of prophets. They add up to a litany that at times sounded like tiny whimpers and at others like thunder which, in the end, translated into the thunder of cannons. Very few listened and one-and-a-half million casualties followed. But as Bruce explains, "The Cassandras, the Jeremiahs, the graffiti at the feast have never been welcome."

The horror that came in 1861 sent before it a long shadow. That is not surprising since the Union itself was born in war. Barely a generation passed after the Revolution before Americans fought a war once again against the British, what Henry Clay called "the second American War of Independence." People were sensitized to war, and the possibility of a civil war occurred to some early and never disappeared thereafter. Internal disputes that sometimes flared into violence conjured up dire visions. This was true of the Shays and the Whiskey rebellions, and of the Kentucky and Virginia resolutions opposing federal authority. The "reassuring misfire" at Hartford, to quote the inimitable Bruce, where some New Englanders tried to raise secessionist sentiments, built on the angst of those locals who could "not abide the thought that Western expansion and population growth might leave that proud section a mere dog-ear on the national map."

Jefferson's "fire-bell in the night" followed—the Missouri Compromise line across the growing United States, making the North free, and the South slave, "a geographical line, coinciding with a marked principle, moral and political." "The knell of the Union," the old Virginian shuddered. But the younger John Quincy Adams, Secretary of State, thought that if a combined servile and civil war could end slavery, "as God shall judge me, I dare not say that it is not to be desired."

The war to come was starting to take shape out of the vague shadow. Still, compromises prevailed in 1820 over Missouri, then ostensibly over the tariff in the nullification crisis of the next decade, over slavery in 1850, and in myriad lesser ways as well. The fear of war, Bruce thinks, "more than any other single factor" led to the continuation of peace even as "generations of lurid warnings had also made civil war a more familiar hence less frightening idea."

Then the shadow ended: the country was about to go to war. And thus Bruce brings us to Abraham Lincoln. His Confederate counterpart, Jefferson Davis, seemed to understand something of the horror that was about to descend upon the land but, always the good soldier, stood ready to do his "duty." Abraham Lincoln on the other hand, like most of his countrymen, North and South, did not have the strength of vision to face reality. When getting a glimpse of the possibility of civil war he relied on the psychological mechanism of avoidance. What he could not handle he would not consciously brave. As a young man in Illinois, Lincoln already refused to believe that a terrible fraternal war could descend upon what he later called "this almost chosen people." Into the late 1850s he continued "his desperate dodging of the war shadow." "There will be no war, no violence," he repeated in awesome error.

When the war came, James McPherson explains, the president still hoped to follow a moderate, conciliatory course, restore the "Union as it was," and do no more. He did not want the conflict to "degenerate into a remorseless revolutionary struggle." Change, he hoped, would come as "the dews of heaven." But that was not to be, and the war turned Lincoln into something of a revolutionary.

McPherson focuses on the war—for what shaped "Lincoln's greatness was the war." Chapter Two, "Lincoln and the Strategy of Unconditional Surrender," borrows Prussian theorist Carl von Clausewitz's distinction between national strategy (a country's war goals) and military strategy (the use of force to reach the goals). Lincoln has been described by more than one historian as a military genius, a master strategist, but McPherson explains that whatever successes and failures the

president had in that sphere, his creation of a politically successful national strategy was his "real strategic contribution . . .
to Union victory." After Vietnam the world needs no reminder
that wars often become muddles of confused purpose. Lincoln
gave purpose to the American Civil War.

To do so as early hopes of an easy victory disappeared, he
adopted a strategy of total war. He would abolish slavery and
change fundamentally the social and political system of the rebels. He would settle for nothing less than unconditional surrender. Black freedom became "an end as well as a means of
Union victory"; in Lincoln's words at Gettysburg, "a new birth
of freedom."

McPherson thus makes emancipation, the central element
of Lincoln's victorious national strategy, the measure of his
greatness. David Brion Davis, in turn, focuses on "The Emancipation Moment" in Chapter Three—that instant of transfiguration in history that turns a slave into a free person. He notes
both the epiphany of a glorious moment and the often disap-
pointing realities that surround the ending of bondage. And
he looks not merely at the United States but at other slave
societies as well: the British West Indies, Cuba, Brazil, Surinam, French colonies, even twentieth-century Africa. In the
process he illuminates both America and its emancipating president.

For a century and more Lincoln has been identified with
an idealized emancipation moment that had its roots in the
Bible, in the ancient Hebrew Jubilee (a day of atonement and
slave liberation) that was given chain-shattering Christian
meanings, especially by slaves. More recently Lincoln's role in
ending bondage in America has been questioned by historians,
and often heated debates have ensued. Davis, by casting a long
and wide glance, creates perspective and, one hopes, some calm.
He thus carefully examines an 1840 French bronze relief by

Pierre-Jean David d'Angers depicting the emancipation moment—rather than the analogous sculpture of Thomas Ball, for example, which shows Lincoln with a slave rising from his knees. It was dedicated in Washington by Frederick Douglass in 1876.

Such works of art, actual ceremonies in the United States on January 1, 1863, indeed the legend of emancipation as a whole, "tried to express the emotions evoked by the idea of an instantaneous deliverance from evil." Davis suggests that the idea was deeply influenced by that of manumission, which he defines as "a voluntary speech act that annihilates a status without altering the status quo." But he is not pinned down easily and also speaks of "execution, baptism, marriage, divorce, or discharge from armed service" as, in some ways, comparable moments of passage. Political leaders often promoted the idea of emancipation as a moment of transfiguration but in practice they worked for the control of the freed people and the maintenance of order and social continuity. Nonetheless violence, or the threat of it, was always a close relative of emancipation proclamations. Abraham Lincoln, Davis suggests, fits right into the pattern of history. The scholar thus finds less of a champion of freedom in Lincoln than, through the contingencies of history, a symbol of it, one that Martin Luther King, Jr., for example, could use forcefully a century later.

In Chapter Four, "One Among Many: The United States and National Unification," Carl Degler, too, attempts to understand "the primal event in American history," and its leading figure, by looking at the rest of the world, especially Europe. He knows that the Civil War is "a peculiarly American event, one central to our national experience," yet he also finds challenging new insights by following the comparative route.

Degler takes his text from the nineteenth-century French historian Ernest Renan: "Deeds of violence . . . have marked the origin of all political formations, even those which have

The American Bismarck? Was Lincoln a
man of blood and iron? This cartoon by
Robert Grossman first appeared in the 1990
New York Times and accompanied an excerpt from Carl Degler's chapter in this
book. *Courtesy of Gabor Boritt.*

been followed by the most beneficial results." Degler proceeds to give an appraising glance to nineteenth-century attempts at nation-building in Hungary, Poland, Italy, Japan, Germany, Switzerland, and Canada. Among these he finds the closest analogy to the United States in the brief Swiss civil war of the 1840s, but the most interesting one in Germany's uniting wars between 1866 and 1871.

From such a perspective Lincoln becomes something of an American Bismarck. "Historians of the United States have not liked comparing" the two, Degler remarks, but he ventures forth resolutely. The German and the American both created nations, though Lincoln's task was greater, hence his means "harsher, more deadly." At Fort Sumter he helped provoke the war, somewhat like Bismarck with his infamous 1870 "Ems dispatch." "Lincoln's nationalism needed a war." Like his European counterpart, he stretched the law of the land in numerous ways, culminating in the Emancipation Proclamation. McPherson argues that Lincoln abolished slavery (apart from his hopes for a free United States) for reasons of strategy, to win the war. Degler goes further, saying that emancipation was "virtually dictated" in the president's mind because slavery was the foundation of an incipient rival nation. In such a portrait Lincoln appears much less likeable than is usually assumed, but in historical terms also the greater, the creator of the modern American nation.

Kenneth Stampp's Chapter Five, "One Alone? The United States and National Self-determination," continues with a question of global importance, but focuses principally on its native manifestation. Americans, he explains, have an age-old tradition of supporting the self-determination of nations around the world. Examples abound, ranging from early nineteenth-century stands encouraging the independence of the Latin American states, to Woodrow Wilson's policies early in the twentieth century proclaimed eloquently in the Fourteen Points. However, tending to see themselves as an exceptional people, one alone, Americans, with the exclusion of some Southerners, do not usually think of their own Civil War as a question of self-determination. All the same, Stampp argues, the one great attempt at applying the principle in the United States came in the Civil War. And "the South learned at a terrible price" the harsh truth of the American double standard. In short, the country had "one tradition for export, another for use at home."

As a young man Lincoln had spoken in favor of self-determination in Hungary and elsewhere, even favored the right of revolution. But by 1861 he sided with a second, more recent, and more questionable American tradition which declared that the Union must be preserved at all cost. What followed was the American bloodbath. In leading the North toward that gory road, Lincoln's "was the decisive voice."

A war for the Union, the oppression of the southern majority, was a morally ambiguous undertaking. Easy triumph eluded both sides and at last Lincoln turned to emancipation. Stampp makes the startling suggestion that even then the president did not fully understand what he had done, that he had transformed the war into a great social revolution. By 1863, however, "the war had gone on too long, its aspect had become too grim, and the escalating casualties were too staggering for a man of Lincoln's sensitivity to discover in that terrible ordeal no greater purpose than the denial of the southern claim of self-determination." Unbearable pain brought understanding. At Gettysburg, Lincoln spoke of the hope that Americans "shall not have died in vain—that this nation, under God, shall have a new birth of freedom." He was well on his way to becoming the Great Emancipator. He would try to make no peace and save the Union without black freedom, not even in the face of the military and expected political defeat of 1864. He then pushed through the Thirteenth Amendment abolishing slavery under the Constitution. And at last, close to the end of the war, and his life, Lincoln's second inaugural address attested that saving the Union had grown for him "ancillary" to the liberation of the slave millions.

The president could not in his time, nor the world since, resolve the issue of who and at what point was entitled to self-determination. As a rule, force more than justice and morality made decisions. But, in the American Civil War, Lincoln escaped the dilemma by "embracing the cause of the slave" and so finding "the war's ultimate justification."

If during the war Stampp's Lincoln grows to greatness, he is met there by Arthur Schlesinger's Chief Magistrate in Chapter Six, "War and the Constitution: Abraham Lincoln and Franklin D. Roosevelt." And like Degler, Schlesinger follows a comparative approach. The sixteenth president and the thirty-second were very different men, but they also shared many similarities. Most important, they both faced epic emergencies, braved being denounced as dictators, and risked setting fearful precedents by moving beyond the fundamental law of the land while making war and protecting internal security.

That at the end of the eighteenth century the framers of the Constitution had put firm checks on executive powers did not prevent "unauthorized presidential adventurism" from thriving in the young republic. From Thomas Jefferson's war against the Barbary pirates the country moved to the Mexican War, denounced by a lanky young congressman from Illinois as unnecessarily and unconstitutionally begun. Schlesinger does not elaborate on Lincoln's role in starting the Civil War, but shows that once war came the president acted with unprecedented vigor. "Must a government . . . be too *strong* for the liberties of its own people, or too *weak* to maintain its own existence?," he asked, and during the next four years proved otherwise. But in the process he spent unauthorized funds and, without consulting Congress, imposed a naval blockade, enlarged the armed forces, and suspended the writ of habeas corpus. Later he imposed martial law and military courts on civilians, allowed untold numbers of arrests, suppressed newspapers, denied the mail to the "treasonous," and seized private property. In short, he called out the war power of the Constitution with a vengeance in however disorganized, *ad hoc* a manner; "the law of war, in time of war," he said.

Of course Lincoln usually got Congress to support him, at least retroactively. In the case of the Emancipation Proclamation, however, he did not permit even his Cabinet much of a voice. Cool-headed Lord Bryce later described him as "almost a dictator . . . who wielded more authority than any single Englishman has done since Oliver Cromwell." More authority? Perhaps. But dictatorship? Very far from it, as Schlesinger shows clearly and concisely.

Four score years later Franklin Delano Roosevelt followed suit. Once again America itself was at stake. Harking back to the Civil War, Roosevelt labeled his opponents "Copperheads," sent U.S. destroyers to aid Winston Churchill's beleaguered Britain, and fought something of an undeclared naval war in the North Atlantic on its behalf—and, as he believed, that of the United States and freedom.

After Pearl Harbor, the president saw internal threats in all too many places. Shameful abuses abounded, with Japanese-Americans as the special targets. And yet, once again on the whole, due process was observed and most ordinary freedoms continued. If anything, one is surprised by how much liberty—from great freedom for the press to nearly normal elections—flourished during those dreadful years.

The greatest danger with both presidents came from their risking the creation of dangerous precedents. But as historian

Henry Adams explained after the Civil War: "If the Constitutional system restored itself, America was right." And thus she turned out to be.

Lincoln defined the predicament faced by the war president this way: " 'Was it possible to lose the nation, and yet preserve the Constitution?" And if that was too abstract, he explained so that no one could misunderstand: "Often a limb must be amputated to save a life; but a life is never wisely given to save a limb." So it was in the Civil War. And so it was in the Second World War.

Schlesinger, the scholar who gave us the pejorative phrase "imperial presidency," admires the records of both Lincoln and Roosevelt. Their trials were fiery; their triumphs all the greater. He even accepts their excesses because the very life of the nation was at stake. As a historian who had served in the White House he knows, better than most scholars, that criticizing historical figures with hindsight is nearly as easy as quarterbacking on Monday morning. He cites Maitland: "It is very difficult to remember that events now in the past were once in the future." And adds: "We know how it all came out. Lincoln did not."

The final chapter of the book, "War Opponent and War President," brings Lincoln full circle. It begins with the influences on his early childhood and closes with the end of his life. The preceding chapters tended to sketch Lincoln with broad lines; this one goes into minute detail but, like the rest of the book, looks for the large picture. The subject is Lincoln's views of war itself.

I begin with a glance at anti-war thought in the Western world from antiquity through the Enlightenment of the eighteenth century. By then the millennia-old feelings crystallized into a liberal belief that war was an aberration. It could only be justified under extreme circumstances, when liberty itself was at stake. What liberty meant, however, was open to varying interpretations.

Of course war itself survived and, in America as well as elsewhere, was often celebrated. As Lincoln looked back at the War of Independence and the War of 1812, he, too, saw its benefits, but he also clearly recognized the horror that war was. The failures of civil discourse, the violence that thrived in Jacksonian America, he learned to loathe as well. His sole military experience, as a young volunteer opposing an Indian rising, reinforced such feelings though he never saw action and his recollections favored stressing the ridiculousness of his own soldiering.

Not surprisingly, when he first went to Washington as a congressman in 1847 and found a favorable atmosphere for denouncing his nation's war against Mexico, Lincoln showed little restraint. Even as the country was creating a new generation of military heroes, he defined "military glory" as "that attractive rainbow, that rises in showers of blood—that serpent's eye, that charms to destroy." And yet an essential and typical Lincolnian ambivalence moderated his passionate words as he voted, for example, like others for war supplies for the men committed by the government far from home. Later, when denounced as a new "Benedict Arnold," he also found it in his heart to blur his war opposition by fudging his own history, and by praising heroes and glorious deeds of war. Was Lincoln a war opponent? As he himself admitted, that depended on definitions.

"The shadow of a coming war," as Robert Bruce shows, Lincoln failed to recognize. That made it easier to stand up unyielding against the expansion of slavery. Still he continued to see "war and violence as failures of democracy," the unforgivable exchange of the ballot for the bullet. It was a peace-minded man who found himself in the White House in 1861. He tried to quell war fever by declaring vainly that war would resolve nothing. But he would not back off on the question of slavery expansion, and when war came he accepted it, not understanding the extent of the horror he had accepted. He learned, and learned also that war does accomplish even epic social revolutions and so became the greatest of America's war leaders. The pacific man fighting for Union and, what it stood for, Liberty, thus provides "a poignant testimony to the liberal dilemma."

He attested to his good faith by rapidly growing old, and in the end his life was added to that of the other casualties.[2] Much of the world has approved Lincoln's road and so approved his "love of peace and fighting of war." He has become a light for our future. But the question must still be asked what kind of light is he in an age of megatons and star wars?

A degree of distortion creeps into this introduction, which weaves together with its focus on Lincoln the work of seven historians. Illuminating him was not always the chief intention of these contributors, though that was a happy achievement of their efforts. In any case, providing an organizing principle will help students of history to digest the book.

Useful, challenging questions might also be asked about each of these seven chapters. Disagreements could be highlighted. But an editor's job is neither to referee nor to argue with his historians, much less with himself. Surely many others will take those routes.

Two sensibilities appear to pervade the book. One is the importance of black freedom to understanding Lincoln and his era. The other is the awfulness of war. "The glory road" of yesteryear has become the gory road for us. These interpretations, too, carry inherent dangers.

In 1830 as the United States embarked on the period that would end in secession and point to war, in words as unforgettable as Lincoln's, Daniel Webster proclaimed "Liberty and Union, now and forever, one and inseparable." These words can serve as touchstones in the attempt to understand the work during the past generation of the historians of the Civil War president. Some see Lincoln as putting the Union above all else. Some see him making Liberty paramount. Readers will be able to fit most of the historians of this book into these two groups too, even when an author takes no explicit stand on the issue.

One final question remains, about Lincoln's importance. We ask here some very difficult questions about his role in American history in his lifetime and down to our own time. When all is said and done, however, we find not merely the central figure of American mythology—as scholars often do—but, as far as historians can admit to such, the central figure of American history.

1

The Shadow of a Coming War

ROBERT V. BRUCE

I TAKE MY TEXT from the eighteenth-century poet Thomas Campbell, who won his niche in Bartlett's *Familiar Quotations* with the line "coming events cast their shadows before." That may not always be true of horse races, earthquakes, and the stock market, but it is usually true of wars, at least major ones. And of all American wars, the Civil War cast the longest shadow in advance, far longer than most of us realize. How did the foreshadowing affect the ultimate event? In all our decisions, including those that set the course of history, we act not only on our perceptions of the past but also on our expectations of the future. In the case of the Civil War, what were those expectations? What aroused them, and when, and why? How did they affect choices? Were they used to serve conscious ends? If so, what ends, and how? Could they also have had a subconscious influence?

To begin with, why *was* the shadow of the coming Civil

War so long and so evident? Not all shadows of coming war are plainly visible to those on whom they fall. That of World War II was an exception, because the enormous disaster of World War I had sensitized observers to it. Prophets of the Civil War seemed also to be especially sensitized to the danger. Perhaps that sensitivity likewise reflected a trial just past, that of creating and defending the Union thus menaced. For what might be called the Civil War foresight saga began with the Union itself.

The new nation, conceived in liberty, was scarcely two years old when John Witherspoon, a New Jersey congressman, warned that liberty alone was not enough. Without union, he said, liberty might mean only "a more lasting war, a more unnatural, more bloody, and much more hopeless war, among the colonies themselves."[1] Looking back through the battle fog of the bloody Civil War that did come four score and three years later, we might suppose that Witherspoon envisioned a conflict between North and South. But Witherspoon emphatically insisted that he had in mind no specific *casus belli*, no particular person or area.[2]

Doubtless he reasoned by historical analogy. The politically fragmented European continent had a past of repeated wars among the jostling fragments. Presumably, therefore, a fragmented North America would face a similar future (and as it was, we eventually went to war with both our next-door neighbors). That nightmare of Balkanized free-for-alls lived on in American minds, including Abraham Lincoln's, and so became a factor, perhaps a significant one, in the calculations that determined the ultimate crisis.

Thus in 1787 Shays' Rebellion in Massachusetts at once prompted a Connecticut poet to write, somewhat prophetically, "Shall lordly Hudson part contending powers? And broad Potomac lave two hostile shores?"[3] The Shays affair also spurred an ongoing movement toward a constitution "to form a more perfect Union" and "insure domestic tranquility." Later that year, arguing for ratification of the new Constitution, Alexander Hamilton asserted in the Federalist Papers that if the states parted company, "a man must be far gone in Utopian speculations, who can seriously doubt that the subdivisions into which they might be thrown, would have frequent and violent contests with each other." And as evidence he pointed to the history of ancient Greece and modern Europe.[4]

By the end of the century, those who promoted the Union by holding up the dread alternative of war scarcely needed to spell out the connection between disunion and bloodshed. It occurred to eighteen-year-old Daniel Webster early in 1800 when he heard of the death of George Washington, whom he called "the great political cement" of the Union. That unsettling news, along with current political bickerings, moved him to write a friend, "I already see in my imagination, the time when the banners of civil war shall be unfurled; . . . and when American blood shall be made to flow in rivers by American swords!"[5]

A staunchly Federalist senator, Gouverneur Morris of New York, likewise took it for granted in 1802 that disunion meant war. In the course of debate on repeal of the 1801 Judiciary Act, a senator from Virginia happened to remark, "Whatever may be my opinion of the Constitution, I hold myself bound to respect it." It took only that offhand comment to set Morris off. "I, sir," he retorted, "wish to support this Constitution because I love it; and I love it, because I consider it the bond of our union; because in my soul I believe . . . that without it we should soon be plunged in all the horrors of civil war; that this country would be deluged with the blood of its inhabitants, and a brother's hand raised against the bosom of a brother."[6]

Witherspoon and Hamilton, Webster and Morris, spoke of

disunion as a hypothetical case rather than a clear and present danger. If it had remained hypothetical, perhaps the fear of civil war would have faded to nothing. But by the beginning of the new century, the fear began to attach itself to particular issues. As early as 1790 the cantankerous Senator William Maclay of Pennsylvania caught a whiff of what might be coming. When South Carolina chose formally to instruct its representatives on some point, Maclay wrote in his diary, "It is clearly better to give the state an early legislative negative than finally let her use a practical one which would go to the dissolution of the Union."[7] That was a procedural scenario, a preview of nullification. Real confrontations soon followed.

In 1794 the nation had a taste of sectional conflict. Though far from a bloodbath, the Whiskey Rebellion was the largest single instance of armed resistance to federal authority before the Civil War. It was more than an in-state disturbance like that led by the Shays. The anti-tax uprising spread from Pennsylvania through a considerable segment of the frontier, affecting at least twenty trans-Appalachian counties in four states, as well as the Northwest Territory. The sectional threat of civil war passed with only scattered violence, though President Washington had called out 13,000 troops, and Alexander Hamilton had looked forward eagerly to military fame.[8]

Four years later Maclay's bugaboo of nullification was summoned up more elaborately and concretely by the Jeffersonians for use against the Federalists' Alien and Sedition Acts. The legislature of Kentucky adopted resolutions, drawn up by Jefferson himself, that characterized the Constitution as a mere voluntary agreement between sovereign states, which might therefore rightfully nullify any federal laws they deemed in violation of it. Virginia passed somewhat more circumspect resolutions written by Madison. Virginia was already building an armory at Richmond and arming its militia, for the purpose,

John Randolph said later, of meeting any attack by federal troops. Still hankering for military glory, Hamilton was in fact ready to take on the Virginians, but Jefferson used his influence to cool off his side, every state that replied to the Kentucky and Virginia Resolutions hotly rejected them, and the obnoxious Alien and Sedition Acts were allowed to expire peacefully soon afterward.[9]

Thus, besides the collision of independent states after peaceful separation, two other paths to civil war had been marked out: spontaneous uprisings against federal authority, and state nullification of federal law, followed by federal coercion. One more occasion for civil war remained to be offered: the secession of a state or states in defiance of federal military force. It was one of history's numerous ironies that some of the first faint mutterings of state secession arose in New England, which in the final showdown of 1861 would be first to do battle for the Union.

As the nineteenth century began, the Federalist party in New England could not abide the thought that Western expansion and population growth might leave that proud section a mere dog-ear on the national map. The political triumph of the Jeffersonians in 1801, the Louisiana Purchase in 1803, the Embargo in 1808, and the War of 1812 made the New Englanders tremble not only for themselves but also for the morals, leadership, and foreign policy of the nation.[10]

Most Massachusetts Federalists remained faithful to the Union, at least so long as there were no head-on collisions between the federal government and their beloved commonwealth. Should such a case arise, they looked not to outright secession but to interposition or nullification, on the theory of the very Kentucky and Virginia Resolutions they had earlier condemned. But a small circle of extremists, like that of Ruffin, Rhett, and Yancey in the antebellum South, privately talked secession. One of them went public before the House of Representatives in 1811, warning that were Louisiana to be admitted to statehood "it will be the duty of some [states] to prepare definitely for a separation; amicably if they can, violently if they must." For obvious reasons, however, most of them, like their later counterparts in the South, played down the risk of armed conflict.[11]

The furtive movement sputtered for a while in the dark and then fizzled out. It lacked popular support, and some feared that a secession movement would stir up civil war within New England itself. The Embargo was repealed in 1809, the war with England muddled through to a rousing finale at New Orleans, politicians bethought themselves of careers on the national level, businessmen of investments. There were disturbing rumors that the national government planned to mount a military offensive against any secession. In 1815 the notorious and much maligned Hartford Convention, firmly controlled by the moderate party leaders, reported no current grounds for "direct and open resistance," rejected disunion unless the conflict were "radical and permanent" and then only with the consent of all the states of the Union, and reserved even interposition only for more flagrant infractions of the Constitution than had yet occurred.[12]

The rumblings of armed rebellion, nullification, and secession thus far had given definition and credibility to fears of civil war. They had suggested ways in which it might come to pass. But they had not escalated to the cataclysm that haunted the American imagination. In the words of the Hartford Convention, the conflicts were not "radical and permanent." The very horror with which disunion was contemplated showed how strong and deep-seated was the craving for union and national greatness.

The ultimate test, however, would be an issue that would

not go away or be outgrown, that weighed on a formidable section of the Union, that enlisted a cohesive and organized body of malcontents, that permeated their mores and social structure, that aroused basic instincts and deep emotions on both sides. That way lay Armageddon. And the issue was there in slavery based on race. An uncanny, one might almost think malevolent, concatenation of circumstances brought it to a head not long after the reassuring misfire at Hartford.

Emancipation by then was complete or far advanced in the Northern states, reinforced by the political philosophy of the Revolution, and by the moral conviction that flowers in those with no money at stake. Concurrently, stimulated by the cotton gin and rising world demand for its output, slavery had taken root and spread like kudzu throughout the South and Southwest, reinforced by its serviceability to white prejudice against a numerous and growing race that could be neither ignored nor expelled, and by the moral rationalizing inspired in those with a lot of money at stake. The North was dazzled by a vision of unprecedented growth on which a slave-bound South would be a drag. The South had a vision of cotton wealth on which the North was a parasite. In pursuit of these irresistible but incompatible visions, both parties charged westward and collided in Missouri.

By 1819 the free states had a majority in the House of Representatives and a one-state margin in the Senate. But the imminent admission of Alabama as a slave state would tie the Senate score, and that of Missouri Territory, where slavery had a firm foothold, would break the tie in the South's favor, as well as opening the gateway to the West for slavery. The House amended the Missouri bill to bar further entry of slaves; the Senate rejected the amendment. That impasse between North and South, freedom and slavery, at once and for the future transformed the shadowy, inconstant specter of civil war into a solid presence, starkly defined and stubbornly persistent.

As debate dragged on, House Speaker Henry Clay noted with dismay that the words "disunion" and "civil war" were used freely and familiarly, almost without emotion, in private circles.[13] The public rhetoric of Southern congressmen ran red with blood. A Georgia representative charged the North with having "kindled a fire which . . . seas of blood can only extinguish." One of Georgia's senators envisioned "a brother's sword crimsoned with a brother's blood." The other predicted that "the firebrand . . . will require blood for its quenching." The process of transition from stalemate to slaughter was not delineated, but it presumably involved federal military action against secession. Clay himself occasionally threatened to raise troops in his home state of Kentucky to defend Missouri's rights, though his threats were obviously meant to prod the North toward compromise.[14]

Some Northerners were ready to pick up the gage of battle. A public meeting in Ohio declared "the horrors of war" preferable to the spread of slavery.[15] Secretary of State John Quincy Adams wrote privately that "a dissolution of the Union for the cause of slavery would be followed by a servile war . . . combined with a war between the two severed parts of the Union." Since it would result in "the extirpation of slavery from this whole continent," he added, "as God shall judge me, I dare not say that it is not to be desired."[16]

Southern threats paid off, at least in the short run. In return for the admission of Maine and the barring of slavery from Louisiana Purchase territory north of 36° 30′, Missouri was voted in as a slave state. The fourteen Northern congressmen who supported the compromise gave various reasons then and later, but all included fear of civil war. Most of the renegades lost their seats, but war had been postponed.[17] Nevertheless, the threat of civil war had proved useful and so would soon be heard again.

That threat had more effect on politicians than on the general public, both North and South. Popular indifference to what one newspaper called the "windy debates" over Missouri seems strange in hindsight. But the public was distracted by the Panic of 1819 and the consequent depression.[18] And there was a more fundamental reason for its heedlessness. We all know that reason, because it governs all of us. The Cassandras, the Jeremiahs, the graffiti at the feast have never been welcome. What deafens the public ear to them might in psychological terms be called avoidance or denial. What you don't know—or choose not to know—won't hurt you. Such is the folk remedy for fear and despair, age-old but still universal. Through the antebellum years most Americans, including Abraham Lincoln, would look away from the shadow of war until the substance was upon them.

At least one American in 1820 gazed unblinking into the abyss, however. Thomas Jefferson at seventy-seven had lost much of his hope and therefore his zeal for the end of slavery. He opposed restricting it in Missouri. But its sanction there, and the accompanying compromise, did not relieve his mind. He still thought war between the states highly probable. From Monticello he wrote a Massachusetts correspondent what is arguably his most famous private letter, eloquent, elegiac, and chillingly prophetic: "This momentous question, like a fire-bell in the night, awakened and filled me with terror. I considered it at once as the knell of the Union. It is hushed, indeed, for the moment. But this is a reprieve only, not a final sentence. A geographical line, coinciding with a marked principle, moral and political, once conceived and held up to the angry passions

Fire-bell in the Night. The man who heard "the knell of the Union" as early as 1820. Lithograph of Thomas Jefferson attributed to Nicholas Maurin, 1828. The bust of Benjamin Franklin looks on. *Courtesy of the Library of Congress.*

of men, will never be obliterated; and every new irritation will mark it deeper and deeper."[19]

In 1822 Denmark Vesey's abortive slave conspiracy in South Carolina fired up Southern paranoia about threats to slavery and its offspring, Southern culture. As the saying goes, even paranoiacs have enemies. Banished from the South, the anti-slavery movement found a grudging sanctuary in the North. Meanwhile, still hesitant to fight openly under the banner of slavery, the South Carolinians transferred their fear and fury to the tariff issue as (they hoped) a less morally provocative test of nullification. Committed to an export economy, they damned the protective tariff of 1828 as fattening the purses of Northern manufacturers at their expense. The South Carolina legislature promptly declared the tariff act unconstitutional and endorsed John Calhoun's anonymous disquisition maintaining the legality of nullification.

Unionists still lived in South Carolina. That fall one of them dared to rise up at a public dinner for a leading nullifier and remind the feasters of the shadow upon them. "War, immediate or ultimate, must be the result of disunion," he said. "Civil wars are the most bloody, relentless and revengeful that ever shook the earth . . . [and] a dissolution of this union will be succeeded by conflicts more rending than them all . . . *Domestic slavery,* as it exists with us, might cause a war between these states which would stand alone in history." But the peace-lover's croakings had no effect, and he himself was killed in a duel four months later.[20] Further north John Quincy Adams sounded the same warning. If one or more states challenge acts of Congress and command "the resistance of their citizens against them . . . what else can be the result but war—civil war?"[21]

In New England, now secure as the leader of American industry, the once restive Federalists had turned union-loving

Whigs. Their champion, Senator Daniel Webster, proclaimed the glory of union and the horror of civil war in his stunning rejoinder to South Carolina's spokesman, Senator Robert Hayne. In words echoed by countless schoolboys of the future, he spoke, as he had when a teenager himself, of "states dissevered, discordant, belligerent . . . a land rent with civil feuds, or drenched, it may be, in fraternal blood," and like John Witherspoon a half-century before, he called instead for "Liberty *and* Union, now and forever, one and inseparable!"[22] If Senator Hayne conveyed this message to his constituents, it did not sway them. Unappeased by a somewhat lower tariff act in 1832, and hoping to legitimate a weapon against federal meddling with slavery in the states, a Nullification Convention called by the legislature late in 1832 declared both tariff acts null and void in South Carolina as of February 1, 1833, and announced that the state would secede if force were used to collect duties.[23]

President Andrew Jackson wrote privately that "nullification leads directly to civil war and bloodshed."[24] In his official proclamation he was less explicit, warning the nullifiers that "on your unhappy State will inevitably fall all the evils of the conflict you force upon the Government of your country."[25] But coming from Andrew Jackson, a hint was a hammerblow. Publicly vowing to enforce the laws, Jackson sent more troops to the Charleston forts and General Winfield Scott to direct preparations. Robert Hayne, now governor, raised 25,000 volunteers and a 2000-man mounted brigade, and his agents bought 100,000 arms in the North.[26]

The darkening shadow had its effect. To the popular mind, nullification now meant bloody civil war.[27] That arch-nullifier and tub-thumper for secession Robert Barnwell Rhett ranted that if "in the madness of tyranny . . . the fire and sword of war are to be brought to our dwellings, why then, Sir, I say, let them come!" But his followers hastily reined him in and tried to downplay the likelihood of armed conflict.[28] Calhoun and Hayne themselves feared disunion. Calhoun, indeed, had put nullification forward as a substitute. Frightened by the looming of war, he wrote privately in late January 1833, "We must not think of secession, but in the last extremity."[29] Other Southern states rejected nullification. The Georgia legislature, for example, condemned it "as tending to civil commotion and disunion."[30]

Both sides were at least privately relieved, therefore, when Henry Clay, the Great Pacificator, was persuaded to sponsor a compromise tariff. "We want no war, above all, no civil war, no family strife," he told the Senate and the nation. "We want no sacked cities, no desolated fields, no smoking ruins, no streams of American blood shed by American arms!"[31] Taking note of popular qualms about the use of military force, Jackson dropped his threats and signed.[32] But the settlement of 1833 was only another reprieve.

The shape of war to come had grown clearer. After 1833 no issue brought a Southern state into direct conflict with federal authority, aside from a brief spat over the Texas-New Mexico boundary. The South's driving fear was of direct federal interference with slavery within the states. But that, being clearly unconstitutional, was never threatened, except by John Quincy Adams' comment in 1842 that a civil war would itself warrant emancipation as a military measure.[33] As occasions for civil war, therefore, nullification and nongovernmental uprisings were now obsolete. That left secession, motivated simply by apprehension of some ill-defined future threat. The old notion persisted that even peaceful secession would eventually, in the mere course of human cussedness, be followed by war between the segments. William Seward expressed that idea in an 1844 speech.[34] But the dominant war scenario now became one of state secession met by federal force.

The chief irritant during the 1830s and early 1840s was abolitionist agitation. Though that was no doing of the proslavery federal government, a North Carolina congressman in 1841 warned nevertheless that it would lead to disunion and "the horrors of civil war." Henry Clay in 1842 charged that the abolitionists aimed "through blood, devastation, and conflagration to march forward to emancipation."[35]

Unlike abolitionism the proposed Wilmot Proviso of 1846 involved federal action to bar slavery from territory acquired by the Mexican War. Though it did not bear directly on any Southern state and was never adopted, the mere proposal raised the same fear in the South that westward expansion had in Federalist New England, that of being eventually overwhelmed by the growing weight of the West. Moreover, slavery would die if it could not expand, or so the North hoped and the South feared. John Calhoun told the Senate in 1847 that "the day that the balance between the two sections of the country . . . is destroyed, is a day that will not be far removed from . . . civil war."[36] For the first time the South as a whole began considering secession.[37] Edmund Ruffin of Virginia assured the timid that war would not ensue, because the money-grubbing North would not "waste millions in dollars and men to gratify the abolitionists."[38] Robert Rhett said, and apparently believed, that South Carolina could secede peaceably on its own, because the North knew force would raise up a Southern confederacy.[39]

In 1850, as the renewed quarrel over the status of slavery in the newly acquired territories rose to a crescendo, Henry Clay proposed a compromise package and used the fear of civil war as a lever. "War and dissolution of the Union are identical and inevitable," he said, and of all the wars in history "none . . . raged with such violence . . . as will that war."[40] Others echoed him, notably Daniel Webster in his famous "Seventh of March Speech" supporting compromise. "There can be no such thing as a peaceable secession," Webster declared. Disunion "must produce war, and such a war as I will not describe . . . No, Sir! No, Sir! There will be no secession!"[41]

The dying Calhoun's grim demands, by frightening moderate Southerners away from his position, probably won more Senate votes for the Clay package than did the eloquence of Clay and Webster; and in the end it was the tactical brilliance and sub rosa dealings of Stephen Douglas that turned the trick. But in the country at large the words of Clay and Webster were heard and heeded. The shadow of war, more than any other single factor, moved both sides to make their grudging concessions.[42]

But Southern acquiescence was conditioned on the North's full observance of the terms, and some Southerners did not share that hope. Within a year a small South Carolina paper suggestively named the *Southern Republic* declared prophetically that "we will secede . . . The first assault will be made upon us by the Federal Government by the act of retaining the Forts about Charleston. *This will be war.*"[43] Stephen Douglas, who had put through the Compromise of 1850 and called it a "final settlement," undid it in 1854. Blinded by his obsession with opening up the West, he bought Southern assent for his bill organizing Kansas and Nebraska Territories by inserting a clause repealing the Missouri Compromise. That was the final unsettlement. The Kansas-Nebraska Act provoked the founding of a purely Northern, antislavery party, the Republicans, supplanting the Whigs. The struggle between pro-slavery and antislavery settlers for control of Kansas precipitated three years of sporadic violence on the plains and a split within the Democratic party. "Can civil war between North and South be postponed twenty years longer?" an upper-class New Yorker asked in his 1856 diary. "The struggle will be fierce when it comes."[44]

In the 1856 presidential campaign the Republicans called again for the barring of slavery from the territories, thus repudiating the Compromise of 1850. Leaders in several Southern states accordingly threatened secession if the Republican candidate were elected.[45] Convinced that they were bluffing, the Republicans held firm.[46] In Pennsylvania, the Keystone State, the Republicans denied that they wanted disunion or that the South would secede in any event, but the Democrats played effectively on fears of civil war to win the state and the White House with it.[47] The new reprieve was short-lived. On the day after John Brown was hanged for his wild attack on slavery at Harper's Ferry, the *Richmond Examiner* spoke for many Southerners in declaring that "military collision between the North and South" was "inevitable."[48] To detonate it now only required the election of a Republican president in 1860.

It is apparent in hindsight that many of those who reacted to the shadow of civil war did not do so objectively. They had axes—or swords—to grind. Those in both North and South who warned of civil war as a consequence of secession did so to sober up hotheads on both sides and win support for compromise. The Southerners who denied that the North would use force against secession did so to foster secession as an end in itself. The Northerners who insisted that the South would never secede did so to stave off concessions by the North. The consequence of those special pleadings was self-deception, false weights in the balance of judgment, and a tilt to disaster. In the end, those who denied the possibility of war invited it, those who maintained the inevitability of war accepted it.

In his postwar memoirs William T. Sherman raised a question that seems in order here. Pointing out that civil war had been apprehended by most leading statesmen for half a century, he complained that the government had made no military preparation for it. As commanding general for six years before he wrote that, Sherman must have had some notion of what preparation could and should have been made, but he did not divulge it. Some army officers had certainly foreseen the war. Sherman recalled that as early as 1850 Winfield Scott, the commanding general, had told him that civil war was likely.[49] Sherman himself felt the danger during the elections of 1856 and 1860, and Ulysses Grant voted Democratic in 1856 from fear of war.[50] Some other officers misread the omens. Pierre Beauregard, who would authorize the first shot, wrote Sherman in 1860 not to worry about the situation, that the crisis would pass; and Robert E. Lee dreaded secession and wrote even after Lincoln's election, "I hope all will end well."[51]

Who would have made the preparations? Winfield Scott, a Virginian, was commanding general from 1841 to 1861. And for a dozen years before hostilities began, the War Department was headed by Southerners, of whom the ablest and most active was none other than Jefferson Davis of Mississippi during the critical years from 1853 to 1857. Davis fully grasped the likelihood of civil war. Yet he does not seem to have favored the South unduly as Secretary, though he assured a Southern critic that he had not forgotten it. On one occasion he rejected a Georgia man's plea for a government military academy in the South on the grounds that it would tend "to create and increase sectional jealousies."[52] Davis claimed credit as Secretary for an increase in troop strength by four regiments, and as a senator he subsequently urged further modest increases and opposed cuts, at least until after Lincoln's election. But the numbers involved were minuscule by Civil War standards.[53] In any case a significant increase in military installations and equipment would have made little difference when war came, since much of it would have fallen into Confederate hands at the outset.[54] Congressional stinting of prewar repairs on the forts below New Orleans, actually made it easier for Union forces to retake the city after war came.[55] When Secretary John B. Floyd asked for more troops in 1858, an abolitionist senator objected because they might be used against free-soilers in Kansas, and a Georgia senator objected because they might be used against the South if a Republican won in 1860.[56] It is hard to see what Sherman would or could have done if he had been Secretary in the fifties.

States had a clearer field of action. Since its rather ineffective mobilization in 1833, South Carolina had been building up its arms supply and training troops.[57] Other Southern states followed suit after the mid-forties and with special urgency beginning in 1856.[58] As a senator, Jefferson Davis urged an ambitious preparedness program for Mississippi, including armories, factories, and railroad connections. So did Governor Henry Wise for Virginia.[59] John Brown's raid gave the Southern states a good cover story for war preparations, and Lincoln's election further spurred their efforts.[60] But for all the hustle and bustle, the results, measured against what was to come, counted most in nerving the South for its fatal plunge. Looking back after the war, Jefferson Davis remarked on "the entire lack of preparation for war in the South."[61] To the historian, the significance of the militia frenzy is in its revelation of prewar fears.

As in 1820, the fear of civil war was stronger in political leaders than among the people at large. Sherman recalled that "the Northern people generally . . . took no warning of its coming, and would not realize its existence till Fort Sumter was fired on."[62] The psychological defense mechanism of avoidance or denial doubtless accounted for some of that apathy. Furthermore, alarmists had been crying wolf for so long that those who still listened were likely to be either skeptics or fatalists, like people who live on the slopes of volcanoes, in earthquake zones, or, like all of us for more than forty years, under the mushroom shadow of nuclear war.

Generations of lurid warnings had also made civil war a more familiar and hence less frightening idea. Notwithstanding all those years of rhetorical hemorrhaging, it may be doubted that many people comprehended the reality of war, or that anyone can without experiencing it. Few antebellum Americans had such experience. The Revolutionary War had been romanticized and glorified. The Mexican War, likewise romanticized and gloriously profitable to boot, had been relatively short and small-scale. Even Europe had been spared prolonged and profligate slaughter since 1815. Though the Taiping Rebellion killed twenty million in China from 1850 to 1864 and was duly reported in the American press, its swirling armies, so alien and distant, impinged on the American consciousness like so many swarms of gnats, exciting no shiver of self-recognition. Many thousands who by dying would learn better the hardest way of all even thought of war as rather a lark. Blood, it seemed on the eve of war, was the fire-eaters' chaser. At any rate some of them promised to drink all that would be shed. All this suggests why Lincoln could say near the end, "Neither party expected for the war, the magnitude, or the duration, which it has already attained."[63]

As 1860 began, the *Richmond Dispatch,* a Southern unionist paper, tried to convey the danger more concretely than did the traditional vague images of blood and destruction: "A dissolution of the Union . . . would be war from the start, war to the knife and knife to the hilt. The widely extended border between the North and South would be a line of fire and blood. Every accessible bay and inlet of every river would be entered, and, ever and anon, large masses of men hurled upon the capitols and important points of Southern states. But the horrors of ordinary war would be far transcended by the barbarities of this cruel strife." A year later, however, the *Dispatch* itself was whooping it up for secession.[64]

During the crucial presidential campaign of 1860, unionist papers in the South insisted that secession would provoke what one called "a war of extermination."[65] On the other hand, those editors who supported John C. Breckinridge and leaned toward secession denied, of course, that secession would mean war.[66] For their own political ends, Northern papers backing Breckinridge, Douglas, or John Bell warned of secession and probable war if Abraham Lincoln were elected. The pro-Southern *New York Herald* saw the usual "carnage and the flames of war . . . swords dripping with fraternal gore," and so on as fruits of a Republican victory.[67] Most Republicans and their press supporters dismissed all this as an old story and mere bluff.[68] With a less belligerent platform than in 1856, offering tariff protection, homesteads, and other boons along with the territorial plank, they had some success in blurring the slavery issue and hence the shadow of war. The popular mood was actually less frenetic than in 1856.[69]

In the South after Lincoln's election, a growing minority of both unionists and secessionists were resigned to war or welcomed it. Still, many unionist papers in the South, especially in the border slave states, continued to oppose secession in the waning hope of avoiding war.[70] And many secessionists claimed, at least publicly, that the North would not attack seceding states because of Northern public opinion, military unpreparedness, need for cotton, concern for holding the border states, and a Yankee calculation that so expensive an undertaking would not pay.[71] The mirage of Northern acquiescence emboldened South Carolina to secede in December, and the lower tier of slave states soon followed.[72] Early in February 1861 the Confederacy was born, and Jefferson Davis became its first—and last—president. The only remaining question was whether or not the North would fight.

Most Northerners had not believed secession possible; and even after it began, many hoped that war could be averted and perhaps the departed states reclaimed through compromise and sober second thoughts.[73] As for agreeing to Confederate independence, however, most Republicans and Northern Democrats shared the familiar view recapitulated by the *New York Times* in March 1861: "Questions of commerce, of the rights of navigation, of extradition . . . a thousand sources of hostility would be created by the very fact of separation . . . It would be impossible . . . to avoid hostilities for any considerable length of time."[74] By then a recognized Confederacy would have had time to organize and arm. It would have had access to foreign credit and alliances. The Northern public would balk at a long, hard war to conquer and absorb an entire established nation. In this view, therefore, if compromise should fail to prevent secession, only immediate military action would make sense. Even Stephen Douglas, mindful of the Midwest's stake in free use of the Mississippi, had endorsed force as a last resort.[75] And the Republicans had their own contingent of fire-eaters who were less reluctant than Douglas, being confident of a quick victory.[76]

The fate of the nation—or nations—now rested with two men, the opposing presidents, Jefferson Davis and Abraham Lincoln. There is little uncertainty about Davis's reaction to the shadow of war. He had recognized it clearly at least as early as 1849, when he said publicly that if "all other things fail there is left the stern appeal—*to arms.*"[77] Davis knew what war was and what it entailed, having served with distinction in Mexico and as Secretary of War. He dreaded war and also disunion, which he was sure would provoke war. Nevertheless, he said, if the federal government should try to put down secession in a Southern state, most likely South Carolina, "that act of usurpation, folly, and wickedness would enlist every true Southern man for her defense."[78] Davis expressed those views consistently and publicly through the fifties. Less than a month before he became president of the Confederacy he wrote privately, "Civil War has only horror for me, but whatever circumstances demand shall be met as a duty."[79]

The mind of Abraham Lincoln was not so transparent, a fact that gave employment to a horde of journalists in his day—and to generations of historians and biographers since. There may be a clue to his thinking in an address titled "The Perpetuation of Our Political Institutions," which he delivered in 1838 before the Young Men's Lyceum of Springfield, Illinois, when he was about to be twenty-nine. The Lyceum Address has been read to shreds with bizarre and dubious results since Edmund Wilson ruminated on it in 1962. Yet there is a remarkable fact about it that has been little remarked. When Lincoln gave the talk in 1838, the Missouri Crisis, the Nullification Crisis, and the anti-abolitionist uproar were of recent memory. His Whig heroes Clay and Webster had seized the nation's attention with their warnings of a sectional conflict. Yet in a speech devoted to the danger of a violent internal upheaval that would threaten American government and principles, Lincoln made no mention of a possible civil war between North and South. His theme was "savage mobs" and "disregard for law," and he stressed that they pervaded both sections.[80] Since Clay and Webster were not reticent about a possible war between the states, there should have been no reason for Lincoln to have shunned the subject before a small, local, and friendly audience—no reason, that is, except his personal reluctance to face the prospect. In short, he seems to have had a mental block, an avoidance mechanism. Why? Something innate in his temperament? Fond memories of his Southern childhood? Or on the other hand,

Antebellum. The man who did not expect war. Photograph by
Alexander Hesler, 1860. *Courtesy of James Mellon.*

some childhood trauma? The answer is likely to remain another aspect of what Richard Current has called "the Lincoln nobody knows."[81]

At any rate, his desperate dodging of the war shadow is strikingly evident in the 1850s. He knew that if the only choice were between disunion and war, he would choose war. Yet only once, to my knowledge, did he let himself be jolted into saying so. Goaded by Democratic charges in the 1856 campaign that Republican victory would trigger secession, he said in a speech at Galena, Illinois, "The Union . . . won't be dissolved. We don't want to dissolve it, and if you attempt it, *we won't let you*. With the purse and sword, the army and navy and treasury in our hands and at our command you *couldn't do it*." But he caught himself immediately and added, "All this talk about the dissolution of the Union is humbug—nothing but folly." A month later in another speech he betrayed desperation. Would there be secession? "The South do not think so," he said. "I believe it! I believe it! It is a shameful thing that the subject is talked of so much." And in other speeches that summer he repeatedly denied that secession would ever happen.[82]

Lincoln never thereafter admitted publicly or privately that civil war might come to pass until it was almost upon him. In his "House Divided" speech of 1858, he said "I do not expect the Union to be *dissolved*." Lincoln, who chose his words with care, spoke not of mere hope, but of expectation. In the famous debates that followed, Douglas accused Lincoln of inciting civil war. But Lincoln insisted, "There will be no war, no violence."[83] During the presidential campaign of 1860, he said nothing publicly on the subject of war. In a private letter, however, he wrote that the Southern people had too much sense to secede, "at least, so I hope and believe."[84] The newspaperman Donn Piatt, who saw much of him during the campaign, recalled later that "Mr. Lincoln did not believe, could not be made to believe, that the South meant secession and war."[85] Caught between three tragic alternatives—disunion, the perpetuation of slavery, and civil war—Lincoln had apparently willed himself to believe that the unthinkable choice would somehow not have to be made.

He still struggled in the net now and then, hoping, even after South Carolina seceded, that it would come back, hoping, even after the Confederacy was formed, that Southern unionists would prevail and Americans would reunite.[86] He, at any rate, would not strike the first blow. Speaking at numerous places en route to his inauguration, Lincoln told his hearers that "there will be no blood shed unless it be forced upon the Government."[87] And in the Inaugural Address itself he told the South, "In *your* hands, my dissatisfied fellow countrymen, and not in *mine,* is the momentous issue of civil war."[88] A few days before Lincoln's journey to decision, his opposite number, Jefferson Davis, made a similar speaking trip. The difference was revealing. Some listeners thought the North would make concessions and the South would rejoin it. Most thought secession would be final and the North would concede it. Davis told them the North would fight. In Mississippi the governor remarked that the state had arms enough. Davis replied that it would need all it could get and more. "You overrate the risk," said the governor. "I only wish I did," said Davis. He had resigned himself to war.[89]

So, at last, did Lincoln. He rejected the option of letting slavery expand and flourish. "The tug has to come," he said, "and better now than later." Yielding would only encourage further Southern demands. To accept disunion was, as it had always been, intolerable to him. So if the South did not repent, there would be only one terrible course left after all.[90] At this juncture his friend Orville Browning reminded him of the old proposition that two confederacies would eventually come to blows. According to Browning, Lincoln "agreed entirely."[91] The thought may have braced him. At least a choice between war now and a worse one later would be easier than one between war and a permanently peaceful separation. He could also cling to the belief that the war had been "forced upon the Government" and, after four years of it, say in his Second Inaugural Address that one side "would *make* war rather than let the nation survive; and the other would *accept* war rather than let it perish."[92]

Thus the long shadow ended at Fort Sumter. And the war came. So much for nearly a century of dire warnings. One thinks again of Cassandra, whom the gods sentenced always to cry doom and never to be believed.

Today Cassandra would be on a talk show discussing the greenhouse effect. The threat of civil war is long past. A century after the real thing the issue of racism once more pitted the federal government against the Southern states, but there was no stomach for a rematch, no appetite for Civil War II. In the time of Southern white resistance to federally ordered school integration, some Louisiana diehard, perhaps Leander Perez, was said to have urged the antique weapon of state interposition on Governor Earl Long, to which the inimitable Earl of Louisiana replied that the feds now had the atom bomb. That, I think, is as good an epitaph as any for the shadow of civil war.

2

Lincoln and the Strategy of Unconditional Surrender

James M. McPherson

*L*INCOLN CAME TO Gettysburg to dedicate a national cemetery for soldiers who died in what turned out to be the biggest battle in America's biggest war. Terrible and tragic was that war. More Americans, soldiers and civilians, died in it than in all the rest of this country's wars combined. But the Civil War also did more to shape the nation than all those wars combined, except the Revolution. The Civil War preserved from destruction the Union created by the Revolution, while transforming that Union into a different kind of nation—giving it a new birth of freedom, as Lincoln said at Gettysburg, by liberating four million slaves and destroying a social system based on human bondage. An understanding of that war is crucial to a comprehension of American history and to an appreciation of Lincoln's leadership in changing the course of that history.

The most important single circumstance in shaping Lincoln's greatness was the war. He was a *War President*. Indeed, he was the only President in our history whose entire administration was bounded by the parameters of war. The first official document that Lincoln saw as President—at one o'clock in the morning when he returned from the inaugural ball— was a letter from Major Robert Anderson at Fort Sumter stating that unless re-supplied he could hold out only a few more weeks. This news, in effect, struck the first blow of the Civil War, and the fatal shot fired by John Wilkes Booth on April 14, 1865, struck virtually the last blow of the war. During the intervening one thousand, five hundred and three days there was scarcely one in which Lincoln was not preoccupied with the war. Military matters took up more of his time and attention than any other matter, as indicated by the activities chronicled in that fascinating volume, *Lincoln Day by Day*.[1] He spent more time in the War Department telegraph office than anywhere else except the White House itself. During times of crisis, Lincoln frequently stayed at the telegraph office all night reading dispatches from the front, sending dispatches of his own, holding emergency conferences with Secretary of War Edwin Stanton, General-in-Chief Henry W. Halleck, and other officials. He even wrote the first draft of the Emancipation Proclamation in this office while awaiting news from the army.[2]

Lincoln took seriously his constitutional duty as commander in chief of the army and navy. He borrowed books on military strategy from the Library of Congress and burned the midnight oil reading them. No fewer than eleven times he left Washington to visit the Army of the Potomac at the fighting front in Virginia or Maryland, spending a total of forty-two days with the army. Some of the most dramatic events in Lincoln's presidency grew out of his direct intervention in strategic and command decisions. In May 1862, along with Secretary of War Stanton and Secretary of the Treasury Salmon P.

Chase, he visited Union forces at Hampton Roads in Virginia and personally issued orders that led to the occupation of Norfolk. Later that same month, Lincoln haunted the War Department telegraph room almost around the clock for more than a week and fired off a total of fifty telegrams to half a dozen generals to coordinate an attempt to trap and crush Stonewall Jackson's army in the Shenandoah Valley—an attempt that failed partly because Jackson moved too fast but mainly because Union generals, much to Lincoln's disgust, moved too slowly. A couple of months later, Lincoln made the controversial decision to transfer the Army of the Potomac from the Virginia Peninsula southeast of Richmond to northern Virginia covering Washington. And a couple of months later yet, Lincoln finally removed General George B. McClellan from command of this army because McClellan seemed reluctant to fight. A year later, in September 1863, Lincoln was roused from bed at his summer residence in the soldiers' home for a dramatic midnight conference at the War Department, at which he decided to send four divisions from the Army of the Potomac to reinforce General William Rosecrans' besieged army in Chattanooga after it had lost the battle of Chickamauga.

Lincoln subsequently put General Ulysses Grant in command at Chattanooga and then in the spring of 1864 brought him to Washington as the new general in chief. Thereafter, with a commander in charge who had Lincoln's full confidence, the President played a less direct role in command decisions than he had done before. Nevertheless, Lincoln continued to help shape crucial strategic plans and to sustain Grant against pressures from all sides during that dark summer of 1864. In August he wired Grant: "I have seen your despatch expressing your unwillingness to break your hold where you are. Neither am I willing. Hold on with a bull-dog gripe, and chew & choke, as much as possible."[3] When Confederate General Jubal Early drove a small Union army out of the Shenandoah Valley in the summer of 1864, crossed the Potomac, and threatened Washington itself before being driven off, Lincoln went personally to Fort Stevens, part of the Washington defenses, to observe the fighting. It was on this occasion that a Union officer standing a few feet from Lincoln was hit by a Confederate bullet and that another officer—perhaps none other than Oliver Wendell Holmes, Jr.—noticing without recognizing out of the corner of his eye this tall civilian standing on the parapet in the line of fire, said urgently: "Get down, you damn fool, before you get shot!" A chastened President got down.[4]

Grant subsequently sent several divisions from the Army of the Potomac with orders to go after Early's army in the Shenandoah Valley "and follow him to the death." When Lincoln saw these orders he telegraphed Grant: "This, I think, is exactly right." But "it will neither be done nor attempted unless you watch it every day, and hour, and force it."[5] In response to this telegram Grant came to Washington, conferred with Lincoln, and put his most trusted subordinate Philip Sheridan in command of the Union forces in the Shenandoah Valley, where they did indeed follow Early to the death and destroy his army. About the same time, Lincoln approved the plans for Sherman's march through Georgia. It was these three campaigns—Grant's chewing and choking of Lee's army at Petersburg, Sheridan's following of Early to the death in the Shenandoah, and Sherman's march through Georgia and the Carolinas—that finally destroyed the Confederacy and brought about its unconditional surrender.

Commander-in-Chief Lincoln was mainly responsible for this unconditional victory of Union forces. But in the huge body of writing about Lincoln—there are more titles in the English language about Lincoln than about anyone else except Jesus and Shakespeare—only a small number of books and articles focuses primarily on Lincoln as a war leader. In 1982, Mark E. Neely, Jr., completed *The Abraham Lincoln Encyclopedia,* a valuable compendium of information and scholarship—which devotes less than 5 percent of its space to military and related matters. In 1984, Gabor S. Boritt organized a conference at Gettysburg College on recent scholarship about the sixteenth President. This conference had sessions on three books of psychohistory about Lincoln, two sessions on books about his assassination, two sessions on Lincoln's image in photographs and popular prints, one on his economic ideas, one on Lincoln and civil religion, one on his humor, one on his Indian policy, and one on slavery and emancipation—but no session on Lincoln as a military leader. In 1987, the outstanding Lincoln scholar of our time, Don E. Fehrenbacher, published a collection of essays, *Lincoln in Text and Context.* Of its seventeen essays on Lincoln, none dealt with the President as a military leader.[6] I note this not as a criticism of these enterprises, which are superb achievements. Rather, it is a reflection on the nature and direction of modern Lincoln scholarship.

A generation ago, fine studies by two historians named Williams—T. Harry and Kenneth P.—told us everything we might want to know about Lincoln's search for the right military strategy and for the right generals to carry it out.[7] A number of other books and articles has also explored Lincoln's relationships with his generals, the wisdom or lack thereof that the President demonstrated in certain strategic decisions, and a great deal more of a similar nature. Many of these are excellent studies. They provide important and fascinating insights on Lincoln as commander in chief. But as a portrait of Lincoln, the strategist of Union victory, they are incomplete. The focus is too narrow; the larger picture is somehow blurred.

Most of these studies are based on too restricted a definition of strategy. On this matter we can consult with profit the writings of the most influential theorist of war, Carl von Clausewitz. One of Clausewitz's famous maxims defines war as the continuation of state policy by other means—that is, war is an instrument of last resort to achieve a nation's political goals. Using this insight, we can divide our definition of strategy into two parts: First, *national strategy* (or what the British call grand strategy); second, *military strategy* (or what the British call operational strategy). National strategy is the shaping and defining of a nation's political goals in time of war. Military strategy is the use of armed forces to achieve these political goals.[8] Most studies of Lincoln and his generals focus mainly on this second kind of strategy—that is, military or operational strategy. And that is the problem. For it is impossible to understand military strategy without also comprehending national strategy—the political war aims—for which military strategy is merely the instrument. This is true to some degree in all wars; it was especially true of the American Civil War, which was preeminently a *political* war precipitated by a presidential election in the world's most politicized society, fought largely by volunteer soldiers who elected many of their officers and who also helped elect the political leadership that directed the war effort, and in which many of the commanders were appointed for political reasons.

Let us look at this matter of political generals, to illustrate the point that military strategy can be understood only within the larger context of national strategy. Both Abraham Lincoln and Jefferson Davis commissioned generals who had little or no professional training: men like Benjamin Butler, Nathaniel Banks, Carl Schurz, Robert Toombs, Henry Wise, and so on. A good many of these generals proved to be incompetent; some battlefield disasters resulted from their presence in command. Professional army officers bemoaned the prominence of political generals: Henry W. Halleck, for example, commented that

strategy (handwritten margin note)

"it seems but little better than murder to give important commands to such men as Banks, Butler, McClernand, and Lew Wallace, but it seems impossible to prevent it."[9]

A good many military historians have similarly deplored the political generals. They often cited one anecdote to ridicule the process. To satisfy the large German ethnic constituency in the North, Lincoln felt it necessary to appoint a number of German-American generals. Poring over a list of eligible men one day in 1862, the President came across the name of Alexander Schimmelfennig. "The very man!" said Lincoln. When Secretary of War Stanton protested that better-qualified officers were available, the President insisted on Schimmelfennig. "His name," said Lincoln, "will make up for any difference there may be," and he walked away repeating the name Schimmelfennig with a chuckle.[10]

Historians who note that Schimmelfennig turned out to be a mediocre commander miss the point. Their criticism is grounded in a narrow concept of *military* strategy. But Lincoln made this and similar appointments for reasons of *national* strategy. Each of the political generals represented an important ethnic, regional, or political constituency in the North. The support of these constituencies for the war effort was crucial. Democrats, Irish-Americans, many German-Americans, and residents of the watersheds of the Ohio and Missouri rivers had not voted for Lincoln in 1860 and were potential defectors from a war to crush the rebels and coerce the South back into the Union. To mobilize their support for this war, Lincoln had to give them political patronage; a general's commission was one of the highest patronage plums. From the viewpoint of military strategy this may have been inefficient; but from the viewpoint of national strategy it was essential.

And even in the narrower military sense the political patronage system produced great benefits for the North, for without it Ulysses S. Grant and William Tecumseh Sherman might not have gotten their start up the chain of command. Although West Point graduates, both men had resigned from the prewar army and neither was conspicuous at the outbreak of the war. But Sherman happened to be the brother of an influential Republican senator from Ohio and Grant happened to be a friend of an influential Republican congressman from Illinois. These fortuitous political connections got them their initial commissions in the army. The rest is history—but had it not been for the political dictates of national strategy, they might never have been able to make their mark on the history of military strategy.

Clausewitz describes two kinds of national strategy in war. One is the conquest of a certain amount of the enemy's territory or the defense of one's own territory from enemy conquest. The second is the overthrow of the enemy's political system. The first usually means a limited war ended by a negotiated peace. The second usually means a total war ending in unconditional surrender by the loser.[11] These are absolute or ideal types, of course; in the real world some wars are a mixture of both types. In American history most of our wars have been mainly of the first, limited type: the Revolution, which did seek the overthrow of British political power in the thirteen colonies but not elsewhere; the War of 1812; the Mexican War; the Spanish-American War; the Korean War. American goals in World War I were mixed: primarily they involved the limited aims of defending the territory and right of self-government of European nationalities, but in effect this required the overthrow of the German and Austro-Hungarian monarchies. In Vietnam the American goal was mainly the limited one of defending the territory and sovereignty of South Vietnam and its anti-Communist government, but this was mixed with the purpose of overthrowing the political and social system that prevailed in part of South Vietnam and involved attacks on that system in North Vietnam as well.

type 2

World War II and the Civil War were the two genuine examples in American History of Clausewitz's second type of war—total war ending in unconditional surrender and the overthrow of the enemy's political system. By total war I mean not only this absolute war aim but also the total mobilization of a nation's population and resources for a prolonged conflict that ends only when the armed forces and resources of one side are totally exhausted or destroyed.

Common sense, not to mention Clausewitz, will tell us that there must be congruity between national and military strategy. That is, an all-out war to overthrow the enemy requires total mobilization and a military strategy to destroy the enemy's armies, resources, and morale, while a limited war requires a limited military strategy to gain or defend territory. When national and military strategies become inconsistent with each other—when the armed forces adopt or want to adopt an unlimited military strategy to fight a limited war, or vice versa—then a nation fights at cross purposes, with dissension or failure the likely outcome. This can happen when a war that is initially limited in purpose takes on a momentum, a life of its own that carries the participants beyond their original commitment without a proper redefinition of war aims—for example World War I, which became a total war in military strategy without a concomitant redefinition of national strategy and ended in an armistice instead of unconditional surrender. But it produced a peace treaty that Germany resented as a *Diktat* because it treated the Germans as if they had surrendered unconditionally. This in turn generated a stab-in-the-back legend that facilitated the rise of Hitler.

One of the reasons why Allied powers in World War II insisted on unconditional surrender was their determination that this time there must be no armistice, no stab-in-the-back legend, no doubt on the part of the defeated peoples that they had been utterly beaten and their Fascist governments overthrown. The Allies won World War II because they had a clear national strategy and a military strategy in harmony with it—along with the resources to do the job. In the Korean War, disharmony between President Truman, who insisted on a limited war, and General MacArthur, who wanted to fight an unlimited one, resulted in MacArthur's dismissal and a sense of frustration among many Americans who wanted to overthrow the Communist government of North Korea and perhaps of China as well. In Vietnam, the controversy and failure resulted from an inability of the government to define clearly the American national strategy. This inability resulted in turn from deep and bitter divisions in American society over the national purpose in this conflict. Without a clear national strategy to guide them, the armed forces could not develop an effective military strategy.

The Civil War confronted the Union government with these same dangers of unclear national strategy and a consequent confusion of purpose between national and military strategy. Like World War I, the Civil War started out as one kind of war and evolved into something quite different. But in contrast to World War I, the government of the victorious side in the Civil War developed a national strategy to give purpose to a military strategy of total war, and preserved a political majority in support of this national strategy through dark days of defeat, despair, and division. This was the real strategic contribution of Abraham Lincoln to Union victory. His role in shaping a national strategy of unconditional surrender by the Confederacy was more important to the war's outcome than his endless hours at the War Department sending telegrams to generals and devising strategic combinations to defeat Confederate armies.

In one sense, from the beginning the North fought Clausewitz's second type of war—to overthrow the enemy's government—for the Northern war aim was to bring Confederate states back into the Union. But Lincoln waged this war on the legal theory that since secession was unconstitutional, Southern states were still *in* the Union and the Confederate government was not a legitimate government. Lincoln's first war action, the proclamation of April 15th, 1861, calling for 75,000 militia to serve for ninety days, declared that their purpose would be to "suppress . . . combinations too powerful to be suppressed by the ordinary course of judicial proceedings."[12] In other words this was a domestic insurrection, a rebellion by certain lawless citizens, not a war between nations. Throughout the war Lincoln maintained this legal fiction; he never referred to Confederate states or to Confederates, but to rebel states and rebels. Thus, the North fought the war on the theory not of overthrowing an enemy state or even conquering enemy territory, but of suppressing insurrection and restoring authority in its own territory. This national strategy was based on an assumption that a majority of the Southern people were loyal to the Union and that eleven states had been swept into secession by the passions of the moment. Once the United States demonstrated its firmness by regaining control of its forts and other property in the South, those presumed legions of loyal Unionists would regain political control of their states and resume their normal allegiance to the United States. In his first message to Congress, nearly three months after the firing on Fort Sumter, Lincoln questioned "whether there is, to-day, a majority of the legally qualified voters of any State, except perhaps South Carolina, in favor of disunion." And to show that he would temper firmness with restraint, Lincoln promised that while suppressing insurrection the federalized militia would avoid "any devastation, any destruction of, or interference with, property, or any disturbance of peaceful citizens."[13]

This was a national strategy of limited war—very limited, indeed scarcely war at all, but a police action to quell a rather large riot. This limited national strategy required a limited military strategy, so General-in-Chief Winfield Scott—himself a loyal Virginian who shared the government's faith in Southern Unionism—came up with such a strategy, which was soon labeled the Anaconda Plan. This plan called for a blockade of Southern salt-water ports by the navy and a campaign down the Mississippi by a combined army and fresh-water naval task force to split the Confederacy and surround most of it with a blue cordon. Having thus sealed off the rebels from the world, Scott would squeeze them firmly—like an anaconda snake—but with restraint until Southerners came to their senses and returned to the Union.

Lincoln approved this plan, which remained a part of Northern military strategy through the war. But he also yielded to public pressure to invade Virginia, attack the rebel force at Manassas, and capture Richmond before the Confederate Congress met there in July. This went beyond the Anaconda Plan, but was still part of a limited-war strategy to regain United States territory and disperse the illegitimate rebel Congress in order to put down the rebellion within ninety days. But this effort led to the humiliating Union defeat at Bull Run and to an agonizing reappraisal by the North of the war's scope and strategy. It was now clear that this might be a long, hard war requiring more fighting and a greater mobilization of resources than envisioned by the restrained squeezing of the Anaconda Plan. Congress authorized the enlistment of a million three-year volunteers, and by early 1862 nearly 700,000 Northerners as well as more than 300,000 Southerners were under arms. This was no longer a police action to suppress rioters, but a full-scale war.

Its legal character had also changed, by actions of the Lincoln administration itself. Even while insisting that this conflict was a domestic insurrection, Lincoln proclaimed a blockade of Confederate ports. A blockade was recognized by international law as an instrument of war between sovereign nations. Moreover, after first stating an intention to execute captured Southern crewmen of privateers as pirates, the administration backed down when the Confederacy threatened to retaliate by executing Union prisoners of war. Captured privateer crews as well as soldiers became prisoners of war. In 1862 the Union government finally concluded cartel for exchange of war prisoners, another proceeding recognized by international law as a form of agreement between nations at war.

Thus, by 1862 the Lincoln administration had, in effect, conceded that this conflict was a war between belligerent governments each in control of a large amount of territory. Nevertheless, the Northern war aim was still restoration of national authority over the territory controlled by rebels but not the overthrow of their fundamental political or social institutions. This limited-war aim called for a limited military strategy of conquering and occupying territory—Clausewitz's first type of war. In the winter and spring of 1861–62, Union forces enjoyed a great deal of success in this effort. With the help of local Unionists they gained control of western Virginia and detached it from the Confederacy to form the new Union state of West Virginia. The Union navy with army support gained lodgements along the south Atlantic coast from Norfolk to St. Augustine. The navy achieved its most spectacular success with the capture of New Orleans in April 1862 while army troops occupied part of southern Louisiana. Meanwhile, two Union naval forces drove up and down the Mississippi until they gained control of all of it except a 200-mile stretch between Vicksburg, Mississippi and Port Hudson, Louisiana. Union armies under Ulysses Grant and Don Carlos Buell, supported by river gunboats, captured Fort Henry and Fort Donelson, occupied Nashville and most of Tennessee, penetrated far up the Tennessee River into northern Alabama, and defeated a Confederate counterattack in the bloody battle of Shiloh. In May 1862, the large and well-trained Army of the Potomac under George B. McClellan drove Confederates all the way up the Virginia Peninsula to within five miles of Richmond while panic seized the Southern capital and the Confederate government prepared to evacuate it. The war for Southern independence seemed to be on its last legs. The *New York Tribune* proclaimed in May 1862 that "the rebels themselves are panic-stricken, or despondent. It now requires no very far-reaching prophet to predict the end of this struggle."[14]

But the *Tribune* proved to be a poor prophet. The Confederacy picked itself up from the floor and fought back. Guerrilla attacks and cavalry raids in Tennessee and Mississippi struck Union supply bases and transport networks. Stonewall Jackson drove the Federals out of the Shenandoah Valley; Robert E. Lee drove them away from Richmond and off the Peninsula; in the Western theater Vicksburg foiled the initial Union efforts to capture it and open the Mississippi; while Confederate Generals Braxton Bragg and Kirby Smith maneuvered the Yankees out of Tennessee and invaded Kentucky at the same time that Lee smashed them at Second Bull Run and invaded Maryland. In four months Confederate armies had counterpunched so hard that they had Union forces on the ropes. The limited-war strategy of conquering Southern territory clearly would not do the job so long as Confederate armies remained intact and strong.

General Grant was one of the first to recognize this. Before the battle of Shiloh, easy Northern victories at Fort Henry and Fort Donelson and elsewhere in the West had convinced him that the Confederacy was a hollow shell about to collapse. After the rebels had counterattacked and nearly ruined him at Shiloh, however, Grant said that he "gave up all idea of saving the Union except by complete conquest.[15] By complete conquest he meant not merely occupation of territory, but destruction of enemy armies, which thereafter became Grant's chief strategic goal. It became Lincoln's goal too. "Destroy the rebel army," he instructed McClellan before the battle of Antietam, and when McClellan proved unable or unwilling to do so, Lincoln removed him from command. In 1863, Lincoln told General Hooker that *"Lee's* Army, and not *Richmond,* is your true objective point." When Lee again invaded the North, Lincoln instructed Hooker that this "gives you back the chance [to destroy the enemy far from his base] that I thought McClellan lost last fall." When Hooker hesitated and complained, Lincoln replaced him with George Meade, who won the battle of Gettysburg but failed to pursue and attack Lee vigorously as Lincoln implored him to do. "Great God!" said the distraught President when he learned that Meade had let Lee get back across the Potomac without further damage. "Our Army held the war in the hollow of their hand and would not close it."[16] Lincoln did not remove Meade, but brought Grant east to oversee him while leaving Sherman in command in the West. By 1864, Lincoln finally had generals in top commands who believed in destroying enemy armies.

This was a large step toward total war, but it was not the final step. When Grant said that Shiloh convinced him that the rebellion could be crushed only by complete conquest, he added that this included the destruction of any property or other resources used to sustain Confederate armies as well as of those armies themselves. Before Shiloh, wrote Grant in his memoirs, he had been careful "to protect the property of the citizens whose territory was invaded"; afterwards his policy was to "consume everything that could be used to support or supply armies." Grant's principal subordinate in the Western theater was Sherman, whose experience in Tennessee and Mississippi where guerrillas sheltered by the civilian population wreaked havoc behind Union lines convinced him that "we are not only fighting hostile armies, but a hostile people, and must make (them) feel the hard hand of war."[17]

Confiscation of enemy property used in support of war was a recognized belligerent right under international law; by the summer of 1862, Union armies in the South had begun to do this on a large scale. The war had come a long way since Lincoln's initial promise "to avoid any devastation, any destruction of, or interference with, property." Now even civilian property such as crops in the field or livestock in the barn was fair game, since these things could be used to feed Confederate armies. Congress sanctioned this policy with a limited confiscation act in August 1861 and a more sweeping act in July 1862. General-in-Chief Halleck gave shape to the policy in August 1862 with orders to Grant about treatment of Confederate sympathizers in Union-occupied territory. "Handle that class without gloves," Halleck told Grant, and "take their property for public use. . . . It is time that they should begin to feel the presence of the war."[18]

Lincoln also sanctioned this bare-knuckles policy by the summer of 1862. He had come around slowly to such a position, for it did not conform to the original national strategy of slapping rebels on the wrist with one hand while gently beckoning the hosts of Southern Unionists back into the fold with the other. In his message to Congress on December 3, 1861, Lincoln had deprecated radical action against Southern property. "In considering the policy to be adopted for suppressing the insurrection," he said, "I have been anxious and careful that the inevitable conflict for this purpose shall not degenerate into a violent and remorseless revolutionary struggle."[19] But, during the epic campaigns and battles of 1862, the war did become violent and remorseless, and it would soon become revolutionary.

Like Grant, Lincoln lost faith in those illusory Southern Unionists and became convinced that the rebellion could be put down only by complete conquest. To a Southern Unionist and a Northern conservative who complained in July 1862 about the government's seizure of civilian property and suppression of civil liberties in occupied Louisiana, Lincoln replied angrily that those supposed Unionists had had their chance to overcome the rebel faction in Louisiana and had done nothing but grumble about the army's vigorous enforcement of Union authority. "The paralysis—the dead palsy—of the government in this whole struggle," said Lincoln, "is, that this class of men will do nothing for the government, nothing for themselves, except demand that the government shall not strike its open enemies, lest they be struck by accident!" The administration could no longer pursue "a temporizing and forbearing" policy toward the South, said Lincoln. Conservatives and Southerners who did not like the new policy should blame the rebel fire-eaters who started the war. They must understand, said Lincoln sternly, "that they cannot experiment for ten years trying to destroy the government, and if they fail still come back into the Union unhurt." Did they expect the North, Lincoln asked sarcastically, to fight the war "with elder-stalk squirts, charged with rose water?"[20]

This exchange concerned slavery as well as other kinds of Southern property. Slaves were, of course, the South's most valuable and vulnerable form of property. Lincoln's policy toward slavery became a touchstone of the evolution of this conflict from a limited war to restore the old Union to a total war to destroy the Southern social as well as political system.

During 1861, Lincoln reiterated his oft-repeated pledge that he had no intention of interfering with slavery in the states where it already existed. In July of that year Congress endorsed this position by passing the Crittenden-Johnson resolution affirming the purpose of the war to be preservation of the Union and not interference with the "established institutions"—that is, slavery—of the seceded states. Since those states, in the administration's theory, were still legally *in* the Union, they continued to enjoy all their constitutional rights, including slavery.

Abolitionists and radical Republicans who wanted to turn this conflict into a war to abolish slavery expressed a different theory. They maintained that by seceding and making war on the United States, Southern states had forfeited their rights under the Constitution. Radicals pointed out that the blockade and the treatment of captured rebel soldiers as prisoners of war had established the belligerent status of the Confederacy as a power at war with the United States. Thus its slaves could be confiscated as enemy property. The confiscation act passed by Congress in August 1861 did authorize a limited degree of confiscation of slaves who had been employed directly in support of the Confederate military effort.

Two Union generals went even farther than this. In September 1861, John C. Frémont, commander of Union forces in the border slave state of Missouri, proclaimed martial law in the state and declared the slaves of all Confederate sympathizers free. General David Hunter did the same the following spring in the "Department of the South"—the states of South Carolina, Georgia, and Florida where Union forces occupied a few beachheads along the coast.

Lincoln revoked both of these military edicts. He feared

that they would alienate the Southern Unionists he was still cultivating, especially those in the border states of Kentucky, Missouri, and Maryland who had kept their states in the Union but might take them out if the North turned this war for the Union into a war against slavery. Lincoln considered the allegiance of these states crucial; he would like to have God on his side, he reportedly said, but he must have Kentucky, and Frémont's emancipation order would probably "ruin our rather fair prospect for Kentucky" if he let it stand.[21] Moreover, Lincoln at this time was trying to maintain a bipartisan coalition on behalf of the war effort. Nearly half of the Northern people had voted Democratic in 1860. They supported a war for the Union but many of them probably would not support a war against slavery. General McClellan, himself a Democrat as well as the North's most prominent general in 1862, warned Lincoln about this in an unsolicited letter of advice concerning national strategy in July 1862, after he had been driven back from Richmond in the Seven Days battles. "It should not be a war looking to the subjugation of the [Southern] people," the General instructed the President. "Neither confiscation of property . . . (n)or forcible abolition of slavery should be contemplated for a moment. . . . A declaration of radical views, especially upon slavery, will rapidly disintegrate our present armies."[22]

But by this time Lincoln had begun to move precisely in the direction that McClellan advised against. He had concluded that McClellan's conservative counsel on national strategy was of a piece with the general's cautious and unsuccessful military strategy—fighting with "elder-stalk squirts, charged with rose water," as Lincoln put it. By July 1862, when he read this letter from McClellan, Lincoln had made up his mind to issue an emancipation proclamation.

He came to this decision only after a long and frustrating effort to persuade the border states to take the first steps toward

Strategists in Conflict: McClellan and Lincoln. A general remaining solidly conservative *(left, front)* and a president moving toward revolution. Detail from photograph by Alexander Gardner, 1862. *Courtesy of James Mellon.*

gradual emancipation. Lincoln proposed such a program in March 1862. He prevailed on Congress to pass a resolution offering federal financial aid to any state undertaking compensated emancipation. Lincoln thought of this as a means of strengthening the Northern war effort by inducing the border states to join the ranks of free states and thereby to deprive the Confederacy of any hope of attracting these states to their side. Three times in the spring and summer of 1862 Lincoln appealed to the congressmen from border states to endorse this plan. At first he relied on the rhetoric of persuasion. The changes produced by gradual emancipation, he said, "would come gently as the dews of heaven, not rending or wrecking anything. Will you not embrace it?" In May he admonished border-state representatives that Northern pressures for an emancipation policy were increasing. "You can not if you would, be blind to the signs of the times," he said.[23] But they did seem blind to these signs. They questioned the constitutionality of federal aid, objected to its cost, bristled at its veiled threat of government coercion, and deplored the potential race problem they feared would come with a large free black population. By July, Lincoln had moved from gentle persuasion to blunt warnings. He told border-state congressmen that "the unprecedentedly stern facts of our case" called for immediate action. Slaves were taking advantage of the war to free themselves. Tens of thousands had already escaped from their masters and come into Union lines, and this number would soon climb to hundreds of thousands. If the border states did not make "a decision at once to emancipate gradually," Lincoln said, "the institution in your states will be extinguished by (the) friction and abrasion" of war. But they still refused, voting on July 12 by a margin of twenty to nine against the President's proposal.[24]

For Lincoln, this was the last straw. That very evening he made the final decision to issue an emancipation proclamation as a war measure to weaken the enemy. The next day he privately told Secretary of State Seward and Secretary of the Navy Welles of his decision. A week later he announced it formally to the Cabinet. He now believed emancipation to be "a military necessity, absolutely essential to the preservation of the Union," he told them. "The slaves (are) undeniably an element of strength to those who had their service," he went on, "and we must decide whether that element should be with us or against us. . . . We must free the slaves or be ourselves subdued." Lincoln conceded that the loyal slaveholders of border states could not be expected to take the lead in a war measure against *disloyal* slaveholders. "The blow must fall first and foremost on . . . the rebels," he told the Cabinet. They "could not at the same time throw off the Constitution and invoke its aid. . . . Having made war on the Government, they (are) subject to the incidents and calamities of war."[25]

All members of the Cabinet agreed except Montgomery Blair, who objected that this radical measure would alienate the border states and Northern Democrats. Lincoln replied that he had done his best to cajole the border states, but now "we must make the forward movement" without them. They would not like it but they would eventually accept it. As for Northern Democrats, Lincoln was done conciliating them. The best of them, like Secretary of War Stanton, had already come over to the Republicans while the rest formed an obstructive opposition whose "clubs would be used against us take what course we might." No, said Lincoln, it was time for "decisive and extensive measures. . . . We want the army to strike more vigorous blows. The Administration must set an example, and strike at the heart of the rebellion."[26]

We must strike at the heart of the rebellion to inspire the army to strike more vigorous blows. Here we have in a nutshell the rationale for emancipation as a military strategy of total war.

It would weaken the enemy's war effort by disrupting its labor force and augment the Union war effort by converting part of that labor force to a Northern asset. Lincoln adopted Secretary of State Seward's suggestion to postpone issuing the Proclamation until Union forces won a significant victory. After the battle of Antietam, Lincoln issued the preliminary Proclamation warning that on January 1, 1863, he would proclaim freedom for slaves in all states or portions of states then in rebellion against the United States. January 1 came, and with it the Proclamation applying to all or parts of ten Southern states in which, by virtue of his war powers as commander in chief, Lincoln declared all slaves "forever free" as "a fit and necessary measure for suppressing said rebellion."[27]

Democrats bitterly opposed the Proclamation, and the war became thereafter primarily a Republican war instead of a bipartisan war. Some Democrats in the officer corps of the army also complained, and seemed ready to rally around McClellan as a symbol of this opposition. But by January 1863, McClellan was out of the army and several other Democratic generals were also soon removed or reassigned. Most Union soldiers understood and accepted the rationale of emancipation as a military measure to help win the war. To the extent that this measure moved the Democrats toward the position of an antiwar party, it started the process by which an overwhelming majority of soldiers became Republicans. An Indiana colonel put it this way early in 1863: "There is a desire [among my men] to destroy everything that in (any way) gives the rebels strength." Therefore "this army will sustain the emancipation proclamation and enforce it with the bayonet." About the same time General-in-Chief Halleck instructed army commanders that "the character of the war has very much changed within the last year. There is now no possible hope of reconciliation with the rebels. . . . We must conquer the rebels or be conquered by them. . . . Every slave withdrawn from the enemy is the equivalent of a white (soldier withdrawn from) combat." One of Grant's field commanders explained that "the policy is to be terrible on the enemy. I am using negroes all the time for my work as teamsters, and have 1,000 employed."[28]

This military strategy of being "terrible on the enemy" soon went beyond using emancipated slaves as teamsters and the like. Two congressional acts in 1862 had authorized the enlistment of blacks as soldiers, and the Emancipation Proclamation also announced an intention to do so. Implementation of such a truly revolutionary policy of putting arms in the hands of former slaves to fight their former masters proceeded slowly and hesitantly at first. But by early 1863 the administration was moving full speed ahead on this matter. In March, Lincoln wrote to Andrew Johnson, military governor of occupied Tennessee: "The bare sight of fifty thousand armed, and drilled black soldiers on the banks of the Mississippi, would end the rebellion at once. And who doubts that we can present that sight, if we but take hold in earnest?" By August 1863 the Union army had recruited those 50,000, and Lincoln stated in a public letter addressed to dissenting conservatives that "the emancipation policy, and the use of colored troops, constitute the heaviest blow yet dealt to the rebellion."[29]

Emancipation, then, became a crucial part of Northern military strategy, an important means of winning the war. But if it remained merely a *means* it would not be a part of national strategy—that is, of the *purpose* for which the war was being fought. Nor would it meet the criterion that military strategy itself should be consistent with national strategy, for it would be inconsistent to fight a war using the weapon of emancipation to restore a Union that still contained slaves. Lincoln recognized this. Although restoration of the Union remained his first priority, the abolition of slavery became an end as well as a means, a war aim virtually inseparable from Union itself. The first step in making it so came in the Emancipation Proclamation, which Lincoln pronounced "an act of justice" as well as a military necessity. Of course, the border states, along with Tennessee and small enclaves elsewhere in the Confederate states, were not covered by the Proclamation because they were under Union control and not at war with the United States and thus exempt from an Executive action that could legally be based only on the President's war powers. But Lincoln kept up his pressure on the border states to adopt emancipation themselves. With his support, leaders committed to the abolition of slavery gained political power in Maryland and Missouri and pushed through constitutional amendments that abolished slavery in these states before the end of the war.

Lincoln's presidential reconstruction policy, announced in December 1863, offered pardon and amnesty to Southerners who took an oath of allegiance to the Union *and* to all wartime policies concerning slavery and emancipation. Reconstructed governments sponsored by Lincoln in Louisiana, Arkansas, and Tennessee abolished slavery in those states—at least in the portions of them controlled by Union troops—before the war ended. West Virginia came in as a new state in 1863 with a constitution pledged to abolish slavery. And in 1864, Lincoln took the lead in getting the Republican national convention that renominated him to adopt a platform calling for a Thirteenth Amendment to the Constitution prohibiting slavery everywhere in the United States forever. Because slavery was "hostile to the principles of republican government, justice, and national safety," declared the platform, Republicans vowed to accomplish its "utter and complete extirpation from the soil of the republic." Emancipation had thus become an end as well

Unknown Union Soldier. Freeing black people became both a means
to and an end of Lincoln's national strategy. Photographer
unknown. *Courtesy of the Chicago Historical Society.*

as a means of Union victory; as Lincoln put it in his Gettysburg Address, the North fought from 1863 on for "a new birth of freedom." [30]

Most Southerners agreed with Jefferson Davis that emancipation and the Northern enlistment of black soldiers were "the most execrable measures in the history of guilty man." Davis and his Congress announced an intention to execute Union officers captured in states affected by the Emancipation Proclamation as "criminals engaged in inciting servile insurrection." [31] The Confederacy did not carry out this threat, but it did return many captured black soldiers to slavery. And Southern military units did, on several occasions, murder captured black soldiers and their officers instead of taking them prisoner.

Emancipation and the enlistment of slaves as soldiers tremendously increased the stakes in this war, for the South as well as for the North. Southerners vowed to fight "to the last ditch" before yielding to a Yankee nation that could commit such execrable deeds. Gone was any hope of an armistice or a negotiated peace so long as the Lincoln administration was in power; the alternatives were reduced starkly to Southern independence on the one hand or the unconditional surrender of the South on the other.

By midsummer 1864 it looked like the former alternative—Southern independence—was likely to prevail. This was one of the bleakest periods of the war for the North. Its people had watched the beginning of Grant's and Sherman's military campaigns in the spring with high hopes that they would finally crush the rebellion within a month or two. But by July, Grant was bogged down before Petersburg after his army had suffered enormous casualties in a vain effort to hammer Lee into submission, while Sherman seemed similarly stymied in his attempt to capture Atlanta and break up the Confederate army defending it. War weariness and defeatism corroded the morale of Northerners as they contemplated the seemingly endless cost of this war in the lives of their young men. Informal peace negotiations between Horace Greeley and Confederate agents in Canada and between two Northern citizens and Jefferson Davis in Richmond during July succeeded only in eliciting the uncompromising terms of both sides. Lincoln wrote down his terms for Greeley in these words: "The restoration of the Union and abandonment of slavery." Davis made his terms equally clear: "We are fighting for INDEPENDENCE and that, or extermination, we will have."[32] As Lincoln later commented on this exchange, Davis "does not attempt to deceive us. He affords us no excuse to deceive ourselves. He cannot voluntarily reaccept the Union; we cannot voluntarily yield it. Between him and us the issue is distinct, simple, and inflexible. It is an issue which can only be tried by war, and decided by victory."[33]

This was Lincoln's most direct affirmation of unconditional surrender as the *sine qua non* of his national strategy. In it he mentioned Union as the only inflexible issue between North and South, but events in the late summer of 1864 gave Lincoln ample opportunity to demonstrate that he now considered emancipation to be an integral part of that inflexible issue of Union. Northern morale dropped to a low ebb in August. "The people are wild for peace," reported Republican political leaders. Northern Democrats were calling the war a failure and preparing to nominate McClellan on a platform demanding an armistice and peace negotiations. Democratic propagandists had somehow managed to convince their own party faithful, and a good many Republicans as well, that Lincoln's insistence on coupling emancipation with Union was the only stumbling block to peace negotiations, despite Jefferson Davis's insistence that Union itself was the stumbling block. Some Republican leaders put enormous pressure on Lincoln to smoke Davis out on this issue by offering peace with Union as the sole condition. To do so would, of course, give the impression of backing down on emancipation as a war aim.

These pressures filled Lincoln with dismay. The "sole purpose" of the war *was* to restore the Union, he told wavering Republicans. "But no human power can subdue this rebellion without using the Emancipation lever as I have done." More than 100,000 black soldiers were fighting for the Union, and their efforts were crucial to Northern victory. They would not continue fighting if they thought the North intended "to betray them. . . . If they stake their lives for us they must be prompted by the strongest motive. . . . the promise of freedom. And the promise, being made, must be kept. . . . There have been men who have proposed to me to return to slavery the black warriors" who have risked their lives for the Union. "I should be damned in time & in eternity for so doing. The world shall know that I will keep my faith to friends & enemies, come what will."[34]

Nevertheless, Lincoln did waver temporarily in the face of the overwhelming pressure to drop emancipation as a precondition of peace. He drafted a private letter to a Northern Democrat that included this sentence: "If Jefferson Davis wishes . . . to know what I would do if he were to offer peace and re-union, saying nothing about slavery, let him try me." And Lincoln also drafted instructions for Henry Raymond, editor of the *New York Times* and chairman of the Republican national committee, to go to Richmond as a special envoy to propose "that upon the restoration of the Union and the national authority, the war shall cease at once, all remaining questions to be left for adjustment by peaceful modes." But Lincoln did not send the letter and he decided against sending Raymond to Richmond. Even though the President was convinced in August 1864 that he would not be re-elected, he decided that to give the appearance of backing down on emancipation "would be worse than losing the Presidential contest."[35]

In the end, of course, Lincoln achieved a triumphant re-election because Northern spirits soared after Sherman's capture of Atlanta and Sheridan's smashing victories in the Shenandoah Valley during September and October. Soon after the election Sherman began his devastating march from Atlanta to the sea. George Thomas's Union army in Tennessee destroyed John Bell Hood's Confederate Army of Tennessee at the battles of Franklin and Nashville. One disaster followed another for the Confederates during the winter of 1864–65, while Lincoln reiterated his determination to accept no peace short of unconditional surrender. And he left the South in no doubt of that determination. In his message to Congress on December 6, Lincoln cited statistics showing that the Union army and navy were the largest in the world, Northern population was growing, and Northern war production increasing. Union resources, he announced, "are unexhausted, and . . . inexhaustible. . . . We are *gaining* strength, and may, if need be, maintain the contest indefinitely."[36]

This was a chilling message to the South, whose resources were just about exhausted. Once more men of good will on both sides tried to set up peace negotiations to stop the killing. On February 3, 1865, Lincoln himself and Secretary of State Seward met with three high Confederate officials including Vice-President Alexander Stephens on board a Union ship anchored at Hampton Roads, Virginia. During four hours of talks Lincoln budged not an inch from his minimum conditions for peace, which he described as: "1) The restoration of the National authority throughout all the States. 2) No receding by the Executive of the United States, on the Slavery question. . . . 3) No cessation of hostilities short of an end of the war, and the disbanding of all forces hostile to the government." The Confederate commissioners returned home empty-handed, angry because they considered these terms, in their words, "nothing less than unconditional surrender."[37] Of course they were, but Lincoln had never during the past two years given the South any reason to expect otherwise.

The Northern President returned to Washington to prepare his second inaugural address, which ranks in its eloquence and its evocation of the meaning of this war with the Gettysburg Address itself. Reviewing the past four years, Lincoln admitted that neither side had "expected for the war, the magnitude, or the duration, which it has already achieved. Each looked for an easier triumph, and a result less fundamental and astounding." Back in the days when the North looked for an easier triumph, Lincoln might have added, he had pursued a national strategy of limited war for a restoration of the *status quo ante bellum*. But when the chances of an easy triumph disappeared, Lincoln grasped the necessity of adopting a strategy of total war to overthrow the enemy's social and political system.

Whatever flaws historians might find in Lincoln's military strategy, it is hard to find fault with his national strategy. His sense of timing and his sensitivity to the pulse of the Northern people were superb. As he once told a visiting delegation of abolitionists, if he had issued the Emancipation Proclamation six months sooner than he did, "public sentiment would not have sustained it."[38] And if he had waited until six months later, he might have added, it would have come too late.

After steering a skillful course between proslavery Democrats and antislavery Republicans during the first eighteen months of war, Lincoln guided a new majority coalition of Republicans and War Democrats through the uncharted waters of total war and emancipation filled with sharp reefs and rocks, emerging triumphant into a second term on a platform of unconditional surrender that gave the nation a new birth of freedom. Lincoln hoped to achieve a just and lasting peace with malice toward none and charity for all. But until the rebels laid down their arms unconditionally, the war must go on. "Fondly do we hope," said the sixteenth President at the beginning of his second term, "fervently do we pray—that this mighty scourge of war may speedily pass away. Yet if God wills that it continue, until all the wealth piled by the bond-man's two hundred and fifty years of unrequited toil shall be sunk, and until every drop of blood drawn with the lash, shall be paid by another drawn with the sword, as was said three thousand years ago, so it still must be said 'the judgments of the Lord, are true and righteous altogether.' "[39]

3

The Emancipation Moment

DAVID BRION DAVIS

IN 1963 MARTIN LUTHER KING, JR., began his "I Have a Dream" speech at the Lincoln Memorial in Washington by noting that "five score years ago, a great American, in whose symbolic shadow we stand, signed the Emancipation Proclamation." Dr. King's subsequent words, when carefully reviewed, present a puzzling image of the emancipation moment. He spoke of Lincoln's "momentous decree" as "a great beacon light of hope to millions of Negro slaves who had been seared in the flames of withering injustice." The proclamation came, King said, with millennial imagery, "as a joyous daybreak to end the long night of captivity." Yet King immediately and predictably assured his listeners that the Negro, one hundred years later, was still not free. And this magnificent prelude opened the way for his attack on "the manacles of segregation and the chains of discrimination" and for his unforgettable invocation of a dream that as he put it, is "deeply rooted in the American dream," a dream that "one day this nation will rise up and live out the true meaning of its creed."[1]

How could the emancipation proclamation be a beacon light and a joyous daybreak if blacks a hundred years later were still not free? If Lincoln's decree had failed, if America had defaulted on a promissory note and had given the Negro people a bad check, as Dr. King asserted, how could the Lincoln Memorial still be a "hallowed spot" at which civil rights advocates could assemble and remind America of "the fierce urgency of now"? At the end of his speech, after the ringing refrain that dramatized his dream, King quoted an old Negro spiritual to express the climax of the emancipation moment: "Free at last! free at last! Thank God Almighty, we are free at last!" But how could anyone be sure that this sense of fulfillment would be any more effective or permanent than Lincoln's "great beacon light"?

My purpose in raising this question is not to initiate a new round in the perennial debate over the failure of Reconstruction. I am concerned with how the phenomenon of emancipation has been perceived and understood. King's rhetoric highlights the curious fact that Americans have customarily honored the abolition of slavery as a glorious moment of national rebirth and in the next breath have acknowledged that emancipation was in many ways a failure, that its main significance lay in establishing a national commitment that has yet to be fulfilled. No less remarkable is the example of British West Indian emancipation. The sublimity of Parliament's act of 1833 has never been diminished by the knowledge that abolitionists soon discovered that the post-emancipation system of apprenticeship perpetuated most of slavery's worst evils, or that the abolition of apprenticeship was followed by more subtle forms of coercion. Different criteria and even different language are used to describe the epiphany of the emancipation moment and the mundane realities and continuities of the post-emancipation era.

The meaning of emancipation has been profoundly shaped by Judeo-Christian concepts of deliverance and redemption. One thinks immediately of the rich symbolism associated with the Hebrews' deliverance from bondage in Egypt. The promise of God revealing himself to humanity through a chosen people was signified by an emancipation from physical slavery and a grateful acceptance of a higher form of service. Christian commentators frequently elaborated on the significance of the ancient Hebrew Jubilee, the day of atonement and of liberating slaves in the seventh month following seven sabbatical years. The Hebrew Jubilee supposedly prefigured Christ's mission "to proclaim liberty to the captives, and the opening of the prison to them that are bound." These words appear first in Isaiah, but according to Saint Luke, Jesus stood up to read the passage in the synagogue at Nazareth. For many abolitionists and Christianized slaves, these phrases contained the essence of the Gospel message.

To clarify this point, let me quote a characteristic use of the passage in a sermon celebrating the actual moment of British slave emancipation on August 1, 1834. Ralph Wardlaw, a prominent Scottish evangelical, interpreted the Gospel's "proclamation of freedom" to be primarily a "freedom from Sin, from Satan, from Death, from Hell." But by 1833 these terms had become closely identified, at least in the minds of British evangelicals, with West Indian slavery. For Wardlaw the Jubilee occasioned by Parliament's political decision was nothing less than the harbinger of Christ's final salvation of the world: "The trumpet has sounded through all the colonial dependencies of our country, which proclaims 'liberty to the captives.'—O! what heart is there so cold, so seared, so dead, as to feel no thrill of exulting emotion at the thought, that on the morning of this day, eight hundred thousand fellow-men and fellow-subjects, who, during the past night, slept bondmen, awoke freemen!"

Wardlaw expressed confidence that the altruistic act of liberating slaves abroad would restore amity and unity at home and that Britain's example in exerting "a mighty *moral power*" would soon shame the United States and other slave-holding nations "into imitation." British emancipation would thus be "but the first day of a Jubilee year,—of a period of successive triumphs . . . of continuous and rapidly progressive prosperity, to the cause of freedom." In Wardlaw's millennial vision, the total obliteration of slavery would prepare the way for the final Jubilee of universal love and mercy.[2]

Even secular reformers interpreted the British emancipation act as a radical departure from temporizing politics, as the beginning of a new dispensation in which moral principle would triumph over political, racial, and economic conflicts of interest. In commemorating the tenth anniversary of British emancipation, Ralph Waldo Emerson spoke of "an event singular in the history of civilization; a day of reason; of the clear light; of that which makes us better than a flock of birds and beasts: a day, which gave the immense fortification of a fact,—of gross history,—to ethical abstractions." The fact that especially impressed Emerson was that "the negro population was equal in nobleness to the deed." By this he meant that on the night of July 31, 1834, during their last hours as slaves, they had gathered in churches and chapels in order to await their emancipating moment with prayers and tears of joy. There had been no rioting, no feasting or drunkenness, and according to one missionary's report, " 'not a single dance . . . nor so much as a fiddle played.' " The next morning, Emerson assured his listeners, "with very few exceptions, every negro on every plantation was in the field at work." He failed to add that the new apprentices had no other choice. Yet like Wardlaw, Emerson stressed that "other revolutions have been the insurrection of the oppressed; this was the repentance of the tyrant," and the precursor of a new era when "the masses" would awaken and apply an absolute moral standard to every public question.[3]

This idealization of the emancipation moment was most revealingly portrayed in numerous prints and paintings depicting joyous half-clad blacks holding up broken manacles and kneeling in gratitude to well-dressed whites. Here I have time to describe only one example of such iconography, a bronze relief sculptured by Pierre-Jean David d'Angers and dedicated in 1840 at the Place Gutenberg in Strasbourg, France. As part of a series celebrating the history of printing and the power of the printed word, this portrayal of emancipation is appropriately dominated in the center by a printing press from which a large printed sheet emerges. The press has supplanted the darkened stump of a tree, which appears on the far right festooned with chains and a broken manacle. To the left, William Wilberforce, the most famous British abolitionist, rests one hand on the press that served his cause so well. His other hand clasps the back of a naked black male who embraces the liberator with a look of impassioned gratitude. Wilberforce, though dwarfish in real life, towers over the diminutive black whose upturned face displays the grotesque lips and exaggerated prognathism of popular cartoons and caricatures.

To the right of the printing press three naked black men kneel at the feet of Thomas Clarkson, Wilberforce's colleague and occasional rival; Condorcet, the French philosopher and radical theorist who publicized the cause of abolition; and the Abbé Grégoire, the only important French abolitionist who survived the Revolution. Clarkson is untying the ropes that bind the outstretched wrists of the first kneeling figure, who replicates the posture of the kneeling slave in the famous abolitionist icon entitled "Am I Not a Man and a Brother?" The

A European Version of the Emancipation Moment. This dubious idealization by Pierre-Jean David d'Angers, titled "l'Afrique," was dedicated in the Place Gutenberg, in Strasbourg, France, in 1840. *Courtesy of Ville de Strasbourg, Conservation des Musées, Chteau des Rohan.*

posture is also somewhat reminiscent of the portrayal of freed slaves with Lincoln in Currier and Ives prints during the Civil War, for example, and later in sculptures placed in Washington, Boston, and Edinburgh. Traditional European depictions of the "Baptism of the Ethiopian Eunuch" come to mind as well.

In the Strasbourg relief the second black embraces Clarkson's leg, and the third clasps Condorcet's hand. Grégoire is about to open or hand over a book, presumably the Bible. In the background Europeans are distributing books to eagerly raised hands, and to the far left books have been opened for three small naked children who are protected by a uniformed officer. The book imagery may have been inspired by the British and Foreign Bible Society's promise, in 1834, to present

An American Version of the Emancipation Moment. Another dubious idealization, by Thomas Ball, was dedicated in Washington,
D.C.'s Lincoln Park by Frederick Douglass in 1876. A replica was
put up in Boston's Park Square three years later. *Courtesy of the National Park Service.*

copies of the New Testament and Book of Psalms to every freedman who could learn to read by Christmas. Finally, the relief pictures bare-breasted black women cuddling and holding aloft babies who will never know the agonies of bondage.

In actuality, of course, abolitionists were usually far removed from the scenes of emancipation, and the freedmen were well-clad and knelt at no one's feet. But the official ceremonies in slaveholding country tried to express the emotions evoked by the idea of an instantaneous deliverance from evil, a proclamation of liberty to the captives, and the opening of the prison to them that are bound. Like the Strasbourg relief, or the Lincoln prints and sculptures, the emancipation rituals were designed to emphasize the indebtedness and moral obligations of the emancipated slaves as well as their dependence on the culture and expectations of their liberators.

Consider, for example, the living tableau at Camp Saxton, on Port Royal Island, South Carolina, on January 1, 1863. Union forces had invaded and occupied the Sea Islands in November 1861, and the slaves of the region had been consigned to the ambiguous status of contrabands of war. Thirteen months later they were the most significant group of slaves directly affected by Lincoln's Emancipation Proclamation, which applied with a few exceptions to the unoccupied regions of the South still in rebellion against federal authority. Port Royal was the one spot where abolitionists and missionary teachers could celebrate an emancipation moment. Significantly, many slaves refused to attend the ceremony because they feared foul play. But thousands traveled by foot or steamboat, wearing their finest clothes, to hear prayers, hymns, speeches, and the reading of Lincoln's proclamation, which must have been the preliminary Emancipation Proclamation of the preceding September. Neither of Lincoln's proclamations was written to stir the hearts of assembled freedmen (but rather to make them acceptable to often reluctant whites). They are both prosaic documents filled with legalistic whereases, hereafters, thereofs, and to wits, to say nothing of exceptions and qualifications. As English liberals complained, "the principle asserted is not that a human being cannot own another, but that he cannot own him unless he is loyal to the United States." Yet the Sea Island freedmen demonstrated their own loyalty, at the moment when a flag was being presented to the white commander of a new black regiment, by spontaneously singing "My Country, Tis of Thee." Two black soldiers from the regiment then addressed the gathering and called on the freedmen to join in the war to liberate the rest of the South.

Despite the message of immediacy given by emancipation rituals, political leaders worked for gradual, nondisruptive change, not moments of transfiguration. This generalization applies to virtually all the legislative bodies that debated emancipation from the northern state assemblies in the 1780s to the Brazilian Senate and Chamber of Deputies in the 1880s. In retrospect, one is struck by the pervasive fears and extraordinary caution that shaped various proposals for gradual emancipation. Even legislators who believed that slavery was contrary to natural law and sound public policy feared that an uncompensated destruction of one form of property would undermine the trust and expectations of all property holders. They feared that public institutions would become burdened with the cost of supporting children as well as infirm and elderly freedmen once slave owners were relieved of legal responsibility. Above all, they feared that a prolonged period of time would be needed to "prepare" slaves for freedom and to prepare society at large for a difficult transformation. When legislators thought of preparation, they instinctively invoked the model of the growing child, a model considered especially appropriate for so-called primitive or childlike races. The analogy with childhood explains the use of such terms as apprentices and *patrocinados* for adult freedmen. The maturation model also explains the appeal of plans that freed only newborn children and then consigned them to an apprenticeship for twenty-odd years.

As early as 1791, an address to the French National Assembly from the abolitionist Friends of the Blacks expressed the divided mentality that would long influence emancipation plans. In ringing phrases, the reformers affirmed their belief in natural freedom, equality, and inalienable rights and asserted that "the restoration of a slave's freedom is not a gift or an act of charity. It is rather a compelling duty, an act of justice, which simply affirms an existing truth—not an ideal which ought to be." Yet these radical abolitionists also warned the National Assembly that "the immediate emancipation of Negro slaves would be a measure not only fatal for our colonies, but a measure which, since our greed has reduced the blacks to a degraded and impotent state, would be equivalent to abandoning and refusing aid to infants in their cradles or to helpless cripples."[4]

While the legislators of northern states voiced similar concerns, they were mainly interested in finding a way to rid their jurisdictions of the stigma of slavery without incurring financial losses for slaveholders. Of course, the number of slaves in the northern states was very small. In 1790 there were only 3,763 in New England and 36,323 in the Middle Atlantic states. Yet the emancipation laws adopted by the five main slaveholding states freed only the children born after a specified date. For example, slaves born in Connecticut before March 1, 1784, in New York before July 4, 1799, and in New Jersey before July 4, 1804, remained slaves. Although the prospect of mandatory emancipation increased the number of voluntary manumissions, it is clear that opportunistic owners sold many of these hapless blacks to southern traders. The freeborn children worked as unpaid apprentices until they were twenty-five or even twenty-eight, by which time they had compensated their mothers' owners for the financial costs of emancipation. Even so, these laws encountered stiff resistance. In 1817, after a bitter legislative struggle, New York reformers finally secured a measure providing for the total abolition of slavery ten years later. Connecticut took this step only in 1848. New Jersey's abolition law of 1804 was so conservative and slow in operation that the state still contained slaves, euphemistically defined as apprentices, at the time of the Civil War. These facts add perspective to Lincoln's plan of December 1, 1862, to ensure compensated emancipation in all southern states over a period of thirty-seven years.

From the 1780s to the early 1820s British abolitionists sought to liberate West Indian slaves by such a gradual, imperceptible process that there would be no distinct moment of emancipation. The model frequently cited was the slow elevation of European slaves, in the Middle Ages, to the status of serfs and eventually of free peasants. Although the abolitionists envisioned a more rapid transformation than that in the Middle Ages, they wanted at all costs to avoid the kind of bloody revolution that had engulfed the French colony of Saint Domingue, later renamed Haiti. At first they assumed that by cutting off the supply of new slaves from Africa, Britain could induce West Indian planters to take better care of their existing slaves. Enlightened self-interest would then lead to a sequence of reforms that would remove the worst abuses of plantation slavery and finally transform the institution altogether. As they became disillusioned with this original plan, British abolitionists called for more direct administrative and parliamentary pressure to ameliorate the condition of colonial slaves. They adhered, however, to a similar vision of continuous, orderly change. Blacks would be protected from excessive and arbitrary corporal punishment; they would be Christianized and encouraged to form stable family units; they would be granted limited rights to own property and testify in court. Liberty, in short, would be measured out in gradually increasing doses that matched the slaves' preparation. But as George Canning told the House of Commons in 1823, the planters themselves were "the instruments through whom, and by whom, you must act upon the slave population."

The great Jamaican slave insurrection of 1831 suddenly showed that the slaves could act for themselves. Thomas Fowell Buxton, the abolitionists' parliamentary leader, insisted that the slaves ought to be freed instantly, "for I know *our* power of emancipating in one way or another is fast drawing to a close. I mean they will take the work into their own hands."[5] Although various motives contributed to British slave emancipation, government leaders were particularly alarmed by the danger of revolution and increasingly convinced that only drastic reforms could prevent the kind of racial warfare that had ravaged Haiti. Even by 1830, segments of the British public had begun to demand "immediate emancipation," a slogan closely related to the new imperatives of evangelical Protestantism, which had slowly gained support in both Britain and the United States. In America the immediatists hoped for nothing less than a national repentance for the sin of slavery, a cleansing of souls that would impel masters to do their duty. But in Britain immediatists looked to an omnipotent government to issue a decree. For example, in 1830 the veteran abolitionist James Stephen admonished his coworkers to accept no compromise and to propose no measure short of "a general, entire, immediate restitution of the freedom wrongfully withheld." Stephen called on the British public to chant in unison a demand as "simple" as that of Jehovah's messenger to Pharaoh: "LET THE PEOPLE

GO," and then leave the practical means to Parliament.[6] Significantly, government leaders as well as abolitionists accepted this conceptual demarcation between the formal act or command of emancipation, with all its religious overtones, and the "practical" regulations to give the command effect. There was a parallel dichotomy between the "voice" of the British public, interpreted as a pure and spontaneous expression of Christian morality, and the political arts of compromise that were needed to balance contending interests and to advance the common good.

Despite sharp disagreements over the government's apprenticeship plan, virtually all British reformers acknowledged the need to constrain freedmen within a civilizing environment of wholesome discipline, an environment equivalent to that which enlightened masters should have provided had they not shut their minds to amelioration and to their own best interest. The assumptions underlying the abolitionists' own plans and testimony before a House of Lords committee were almost identical to those spelled out by James Stephen the younger, who was legal counsel for the Colonial Office and the government's leading expert on West Indian affairs. "If allowed to do as he liked," Stephen wrote, the freedman "would probably be distracted between the appetites, for basking in the Sun—for pork & Rum—for red waistcoats and the castoff finery of the green room & the servants hall . . . with the occasional addition of a dance or a prayer meeting." According to Stephen, everyone recognized the need for an educational program aimed at Christianizing and civilizing the freedmen, whose aspirations and habits of life should eventually sustain such a demand for the products of human industry "as can be gratified only by persevering and self-denying labor." But since such habits of life would take time to cultivate, especially in the tropics, Stephen argued that ". . . measures must be adopted, tending more directly to counteract the disposition to sloth which may be expected to manifest itself, so soon as the coercive force of the Owners' Authority shall have been withdrawn. The manumitted Slaves must be stimulated to Industry by positive Laws which shall enhance the difficulty of obtaining a mere subsistence."[7]

The laws Stephen had in mind would deny the freedmen access to land and thus rule out alternatives to plantation labor. "The dread of starving," Stephen wrote, "is thus substituted for the dread of being flogged." The fear of starving was what Stephen called "a liberal motive" that marked a transition from "the brutal to the rational predicament." The planter, he added reassuringly, "incurs no other loss than that of finding his whips, stocks and manacles deprived of their use & value."[8]

As it turned out, the government insisted on delaying this transition to "the rational predicament" and on requiring the freedmen to subsidize approximately one-half the cost of their emancipation by working as unpaid apprentices. It was Stephen who drafted the original plan that provided for an apprenticeship of twelve years, which was later reduced to seven years for agricultural workers and five years for artisans and household servants. The apprenticeship plan contained no effective measures to prepare ex-slaves for freedom. It was essentially a substitute for slavery, designed to maintain work discipline and to compensate slave owners, who also received an enormous direct subsidy of twenty million pounds sterling. As a result of mounting public protest and threatened government intervention, the colonial legislatures finally abolished apprenticeship in 1838.

Yet the conservative concerns that had led Britain to adopt apprenticeship did not diminish with the passage of time. The French Orleanist government concluded that the British experiment had failed and contented itself with reports, proposals, and delays until the Revolution of 1848 brought a swift and disruptive end to colonial slavery. In Spain, following a liberal revolution in 1868, Segismundo Moret, a member of the legislature and a vice president of the Spanish Abolitionist Society, reminded the more radical abolitionists that "the great Lincoln did not want to abolish slavery till 1900." Moret's plan, which the Spanish government adopted, emancipated all colonial slaves over the age of sixty and all children born after September 18, 1868. These free-born *patrocinados* were required to work without pay until the age of eighteen and at half-wages until the age of twenty-two. Cubans delayed publication of the Moret law for two years and then made sure enforcement remained in slaveholder hands. In 1880, Spain finally abolished colonial slavery but substituted an eight-year *patronato,* which subjected the apprentices to slave-like discipline and working conditions. The system disintegrated as the *patrocinados* increasingly resisted planter authority; Spain ended the experiment two years before its scheduled expiration date.

Strongly influenced by the example of the American Civil War and Reconstruction, Brazil took an even more cautious approach than Spain's. The law of free birth of 1871 was intended to deprive radicals of an incendiary issue while giving some hope to slaves, who had begun to challenge the labor system. The government rejected the idea of setting a fixed date for ending slavery (one councilor suggested 1930) but adopted an unwieldy and unenforceable program for registering and protecting the free-born children of slaves. By allowing masters to demand and even to sell the "services" of such minors for the first twenty-one years of their lives, the law "respected the past," as the minister of agriculture put it, "and corrected only the future." According to Brazil's most famous abolitionist, the law of free birth led to "another epoch of indifference for the fate of the slave, during which even the government could forget to comply with the law which it had passed." In the mid-1880s a militant abolitionist movement finally encouraged slaves to flee the plantations and provided them protection on trains and in shanty towns around such urban centers as São Paulo. In 1887 the planters of São Paulo province began freeing their slaves, often in exchange for service contracts, as a desperate means of preserving a labor force for the harvest. The mass exodus of blacks created a crisis that in 1888 finally led the government to abolish slavery outright.

During the next two or three decades, the European powers that colonized Africa displayed great caution in interfering with various forms of indigenous bondage. In the regions that they annexed, they generally abolished the legal status of slavery but sanctioned other forms of compulsory labor. Even Britain, which had long campaigned for the total eradication of human bondage, waited until 1928 before freeing the slaves in Sierra Leone, her oldest African colony. In 1925 the report of the Temporary Slavery Commission appointed by the League of Nations vindicated the fears that conservatives had expressed during the previous century-and-a-half: "Wherever slaves have been freed by the stroke of the pen, serious troubles have almost always arisen: an economic crisis caused by diminution of production leading to general impoverishment and even famine, owing to the fact that the freed slaves have regarded their emancipation as meaning the right to do no work; a social crisis, for they sought by irregular and often criminal means to satisfy their daily needs; a political crisis for poverty and disorder made the Government unpopular."[9]

This questionable verdict would have gratified slaveholding planters in the British West Indies, the antebellum South, and mid-nineteenth-century Cuba and Brazil. But the League's so-called experts were not defending slavery. They were social engineers who believed that the institution should be abolished

"progressively"—that is, in a way that maintained the fundamental structure of the previous social order.

From the beginning, I would suggest, the idea of emancipation was profoundly influenced by the model of manumitting individual slaves. Manumission was a ritual of rebirth that
reinforced the existing social order. On this point we are easily
misled by the example of the antebellum South, where slaveholders feared that manumissions would expand a despised
population of free blacks whose presence could only subvert
the system. In most societies, however, slaveowners looked upon
freedmen as allies whose former positions could easily be filled
by the purchase of new captives. They held out freedom as the
reward for a few exceptionally diligent, faithful, or enterprising
bondsmen. Although slavery originates in violence and can always end in a violent overthrow of the master's authority,
manumission is a voluntary speech act that annihilates a status
without altering the status quo. Like the pronouncement of
marriage, it is a speech act that actually effects the thing named.
For example, in ancient India a master would first break a water
jar taken from his slave's shoulder, a symbol of the act of destruction about to occur. He would then sprinkle on the slave's
head parched grain and flowers, symbols of death and life, while
pronouncing three times, "You are no longer a bondsman."
This cleansing ritual should be distinguished from the terrify-
ing capital sentence of English-speaking judges: "that you be
carried from the jail to the place of execution, and there be
hanged by the neck, until you are DEAD! DEAD! DEAD!" Although here the repeated word becomes a simulacrum of the
fate that is ordered, the words themselves do not kill.

Nevertheless, manumission was a moment of passage comparable to an execution, baptism, marriage, divorce, or discharge from armed service. Sometimes the ritual merely legiti-
mated a new relationship that had already evolved, much as

modern marriages and divorces often confirm a process already completed. But usually, manumission marked an abrupt change in the ex-slave's status, identity, rights, and duties. Like various rites of passage, manumission usually required a preparatory period or at least a compensatory payment of some kind. For example, over half the manumissions in Maryland between the Revolution and 1832 were delayed until an adult had worked for a specified number of years or until a child had reached the age of thirty. Like war veterans, discharged convicts, and divorcees, the freedman in most societies carried the stigma of his former status. Indeed, some pre-modern societies regarded slavery as irrevocable as death—a captive slave who had been ransomed by his own people remained the servant or slave of his liberators. And though slaves often took an active role in redeeming themselves, they were defined by the manumission ritual as passive objects, as the recipients of an unmerited gift equivalent to God's gift of everlasting life.

If a slave had earned his own freedom, why should freedom be a gift? As Orlando Patterson has recently explained, bondage was a state of dishonor and "social death." I would also stress that it was associated with cosmic sin, since the Judeo-Christian tradition justified the institution as a punishment for sinful transgression. Because the slave's time and earnings belonged to his owner, no slave could give an equivalent in exchange for the priceless gift of freedom. The slave stood in a position similar to that of the sinner in Calvinist theology. No matter what good deeds or token payments he might offer, freedom was an undeserved gift that created new bonds of limitless obligation. In virtually every slaveholding society, manumission has therefore entailed continuing duties and obligations to the former master and master's family. Instead of being at liberty to do as he pleased, the freedman had acquired the power to serve willingly, as an expression of gratitude. From ancient Rome to colonial Brazil and Surinam, a freedman "client" guilty of ingratitude to his new patron could be re-enslaved.[10]

In the southern states former slaveholders expected similar gratitude—not for the act of emancipation they had been forced to accept, but for the paternalistic care they believed they had given their former slaves. Nothing shattered Southerners more than the discovery that trusted and supposedly loyal blacks felt little obligation toward their former masters. In Maryland, which escaped the constraints of Reconstruction, planters were so incensed that they seized and apprenticed thousands of the freedmen's children as a form of retaliation. But outside forces had destroyed the legitimacy of the slaveholders' authority. It was the abolitionists and government leaders who replaced the manumitting master. As we have seen in our discussion of the Strasbourg bronze relief, the freedmen were supposed to kneel at the abolitionists' feet. Thomas Fowell Buxton was not alone in speaking of "my black clients."

Like manumission, slave emancipation was understood as a ritualized moment of rebirth that ensured a degree of social continuity and of subservience on the part of the freedman. Yet there was a fundamental difference. The Enlightenment and evangelical revivals had discredited slavery itself and had impeached the claims of every slaveholder. Speaking for the Enlightenment, an article in the famous French *Encyclopédie* asserted that since no man could alienate his natural liberty, "the sale of this person is null and void in and of itself: this Negro does not divest himself, indeed cannot under any condition divest himself of his natural rights; he carries them everywhere with him, and he has the right to demand that others allow him to enjoy those rights."[11] For evangelicals who considered slavery the essence of sin, it was as absurd to talk of gradual emancipation as to talk "of gradually leaving off piracy—murder—adultery, or drunkenness." In the words of Andrew

Thomson, a fiery Scottish preacher, slavery was a "pestiferous tree" that had to be cut down and burned root and branch: "You must annihilate it,—annihilate it now,—and annihilate it forever." Such language suggests a total obliteration of the past, an annulment rather than a divorce. Since Thomson considered every hour that men were kept in bondage a repetition of the original sin of man-stealing, he did not shrink from violence: "If there must be violence . . . let it come and rage its little hour, since it is to be succeeded by lasting freedom, and prosperity, and happiness."[12]

The American Civil War was not wholly exceptional in enforcing emancipation with military power. Both in Haiti and the Hispanic American republics, slave emancipation was as much the result of military necessity as of revolutionary ideology. During the Napoleonic wars, Britain weakened her own colonial slave systems by arming and later freeing thousands of West Indian blacks; in 1833 government leaders counted on the loyalty of the free black population and prepared to use force if the planters carried out their threats of resistance. Cuba's Ten Years' War led to the military enlistment of many slaves and to emancipation edicts from the rebels. In short, emancipation decrees were never far removed from violence or the threat of violence. Though often defended as antidotes to slave insurrection, they were also accompanied by warnings of insurrection if the slaves' hopes went unfulfilled. Many British liberals feared that since Lincoln's proclamation applied to regions beyond Union control, it would provoke a bloodbath of racial war.

In 1862, when Lincoln was considering abstract plans for gradual emancipation and colonization, he also knew that the institution of slavery was rapidly disintegrating in the counties along the Potomac. Blacks from Virginia and Maryland were seeking refuge in the District of Columbia and within Union lines, where they brought invaluable intelligence regarding enemy troops, spies, roads, and terrain. Military reverses strengthened the pressure to enlist black troops and to invoke the ultimate weapon of slave emancipation. In the confiscation act of July 17, Congress ruled that the slaves of rebel owners would be "forever free of their servitude" the moment they crossed Union lines. Although Lincoln showed little interest in enforcing this measure, at least until September 22, Congress in effect defined the Union army as an army of liberation. Yet the act applied only to individual fugitives whose owners had engaged in or actively supported the rebellion.

Lincoln ordered the enforcement of this act in his preliminary Emancipation Proclamation of September 22, which also contained the threat that on January 1 he would emancipate the slaves of rebel owners as well as a reference to the plan he would submit to Congress in his annual message of December 1. Contemporaries were struck by the seeming discrepancy between this proposed constitutional amendment of December 1 and the Emancipation Proclamation that Lincoln was scheduled to issue on January 1, 1863. The amendment promised compensation to every state that abolished slavery before January 1, 1900, and federal subsidies for colonizing freedmen outside the United States. The proclamation promised immediate freedom to all slaves in specified rebel territories. For Lincoln, both measures were based on the premise that "Without slavery the rebellion could never have existed; without slavery it could not continue." No less central was Lincoln's conviction "that the people of the south are not more responsible for the original introduction of this property, than are the people of the north; and when it is remembered how unhesitatingly we all use cotton and sugar, and share the profits of dealing in them, it may not be quite safe to say, that the south has been more responsible than the north for its continuance."[13] The proposed amendment and the proclamation were both intended to shorten the war and to reduce casualties and military spending without alienating slaveholders in the crucial border states. But they also exemplified the two images of emancipation that pervaded the nineteenth century—a utilitarian plan attuned to costs, benefits, and population trends and designed to induce slaveholders to act voluntarily in the public interest; and a proclamation of "liberty to the captives, and the opening of the prison to them that are bound."

The various plans for gradual emancipation, including Lincoln's, shared one striking characteristic. They really called for the prolongation of slavery or a slave-like condition of apprenticeship without providing for intermediate states of transition. Lincoln and other planners assumed that the most intelligent and capable slaves would win their freedom at an early stage, but no mechanisms were recommended to identify and protect this talented elite. The prevailing ideal of a laissez-faire meritocracy could never be reconciled with the tendency to think of slaves as an indivisible, corporate group. Moreover, once the ideal of amelioration had been repudiated, little thought was given to guaranteeing slaves an expanding bundle of rights, such as a right to family integrity, property, privacy, leisure time, and self-government within a limited sphere. The possibilities narrowed to two options: first, the postponement to some definite or indefinite date of mandatory but compensated emancipation; and second, the sudden death of what abolitionists like to call a "Satanic institution." The first option would give the nation time, as Lincoln pointed out, to pay off the enormous debt incurred by the destruction of slave property. Lincoln's second option would be punitive to rebel slaveholders, though it would spare southern blacks from up to thirty-seven years of uncompensated labor. Neither course envisioned a transitory process between slavery and freedom.

The nation was tense with uncertainty during the month that elapsed between Lincoln's message to Congress and the day of his scheduled proclamation. In view of the president's repeated statements supporting gradualism, compensation, and colonization, no one could doubt his heartfelt commitment to the sweeping plan he had submitted to Congress. However, many Americans of opposing views did doubt Lincoln's commitment to an unambiguous proclamation. Oliver Johnson summed up the opinion of many abolitionists when he concluded that " 'Old Abe' seems utterly incapable of a really grand action." When Lincoln with a shaking hand finally signed the momentous decree, he was certain he had done the right thing but expressed no jubilation. It was a reluctant act, dictated by the grim necessities of war. Yet despite some grumblings about the geographic limitations of the edict, black and white abolitionists hailed the moment as a true Jubilee or as "the dawn of a new era." Lincoln, wrote one abolitionist, "may yet be the Moses to deliver the oppressed."

One of the most astute assessments appeared in a newspaper in far-off Vienna and was written by none other than Karl Marx. Lincoln, in Marx's view, ". . . always gives the most significant of his acts the most commonplace form . . . Indecisively, against his will, he reluctantly performs the *bravura aria* of his role as though asking pardon for the fact that circumstances are forcing him to 'play the hero.' The most formidable decrees which he hurls at the enemy and which will never lose their historic significance, resemble—as the author intends them to—ordinary summonses sent by one lawyer to another on the opposing side . . . And this is the character the recent Proclamation bears—the most important document of American history since the founding of the Union, a document that breaks away from the old American Constitution—Lincoln's manifesto on the abolition of slavery . . . Never yet has the New World scored a greater victory than in this instance, through its demonstration that, thanks to its political and social organization, ordinary people of good will can carry out tasks which the Old World would have to have a hero to accomplish!"[14]

Marx sensed that Lincoln, precisely because he seemed to be an "ordinary" person, represented a new kind of hero who "in the history of the United States and in the history of humanity" would "occupy a place beside Washington." It may seem strange that a materialist like Marx would attach such importance to mere words, to a decree that lacked the power to free many slaves and that failed to define either the nature or conditions of freedom.

Yet for materialists and idealists alike, Lincoln had pronounced words that transcended the immediate historic moment. No other nation could honor a single Great Emancipator, whose words were analogous to those of a benign master who frees his slaves before suffering a martyr's death. Unlike the details of specific emancipation plans, the context and even content of Lincoln's words did not really matter. They would soon be forgotten. What mattered was the symbolic emancipation moment. It was an enduring moment of promise that would inspire new emancipators like Martin Luther King, Jr. It was a moment that would acquire new meaning as men and women continued to discover new forms of oppression and human bondage.

4

One Among Many: The United States and National Unification

CARL N. DEGLER

*G*IVEN THE COMPELLING interest which the American Civil War inevitably evokes, items in today's newspapers are hardly necessary to attract attention to still another discussion of that primal event in American history. Yet the past few years have undoubtedly awakened more memories of that war and the circumstances that precipitated it than any in over a century. At the fringes of Europe we have seen the secession of the constituent members of a powerful state; to the north of us, a large and self-consciously distinct province, which must remind us in some ways of our own antebellum South in 1860, threatens to leave its Union unless its special character is constitutionally recognized. People have asked, why Mikhail Gorbachev would not be another Lincoln in resisting by force the secession from the Soviet Union, or why hasn't Prime Minister Brian Mulroney of Canada followed Lincoln in promising to enforce the laws if Quebec should carry out its threat to separate from the Canadian Union?

The 1990s, however, have not been the first time in which the American experience has been but one among many. More than a century ago, between the years 1845 and 1870, the world witnessed a widespread efflorescence of nation-building, in the midst of which was the American Civil War. Some of those instances of people's seeking national identity and statehood remind us of the Confederacy inasmuch as they failed to achieve independence. The revolt of the Hungarians against their Austrian masters in 1848 under the leadership of Louis Kossuth was one such failure, though within two decades Hungarian nationalism achieved a kind of acknowledgment of national identity in the dual empire of Austria-Hungary. A more crushing failure was the experience of the Poles who rose in 1863 against their Russian rulers, at the very same time that the United States was struggling to suppress its own uprising in the South. Contrary to the Confederacy's fate, the Polish defeat would be reversed at the end of the First World War.

Other instances of nation-building achieved their aims. In 1847 the Swiss cantons concluded their war for a Union under a new federal constitution and with a fresh and enduring sense of nationality. In 1860, Camillo Cavour of the kingdom of Sardinia with the assistance of France and the military help of Giuseppe Garibaldi brought into being the first united Italy since the days of ancient Rome. During that same decade of the 1860s a united Germany came into existence for the first time as well. Nor were the nationalist outbursts of that quarter-century confined to Europe. They also erupted in Asia, where a new Japan emerged in the course of the Meiji Restoration, in which feudal power was forever subordinated to a centralized state that deliberately modeled itself after the nation-states of Europe.

Looking at the American Civil War in the context of contemporary efforts to establish national identity has the advantage of moving us beyond the often complacent concern with ourselves that I sometimes fear is the bane of United States historians. The Civil War is undoubtedly a peculiarly American event, one central to our national experience. In its endurance, the magnitude of its killing, and the immense extent of its arena it easily dwarfs any other nationalist struggle of its century. Yet if we recognize its similarity to other examples of nation-building of that time we may obtain fresh insights into its character and its meaning, then and now.

First of all, let me clear the ground by narrowing our comparisons. Although the European and Asian instances of nation-building in the years between 1845 and 1870 are comparable to the American experience in that they all involve the creation or the attempt to create a national state, not all of them are comparable on more than that general level. The Meiji Restoration, for example, was certainly the beginning of the modern Japanese state but the analogy stops there since it did not involve a military struggle. The Polish and Hungarian national uprisings bear closer comparison to the Confederate strike for independence, but the differences in nationality between the oppressors and the oppressed (Austrians versus Hungarians; Russians against Poles) render dubious any further analogy to the Confederacy. After all, both the Hungarians and the Poles had been conquered by foreigners; each enjoyed a national history that stretched deep into the past, something totally missing from the South's urge to separate from the United States.

The Italian experience in nation-building comes closer to the North's effort to preserve the Union. A united Italy did emerge eventually from the wars of the Risorgimento and Garibaldi's conquests of Sicily and the Kingdom of Naples. Pertinent, too, is that Garibaldi, as an internationally recognized hero of Italian unification, was entreated by the Lincoln Administration to become a leading officer in the Union army.[1] Yet, neither event offers much basis for comparison. The unification of all of Italy was, as English statesman William Gladstone remarked, "among the greatest marvels of our time," and simultaneously a kind of accident.[2]

It was a marvel because Italy's diversity in economy, language, culture, and society between North and South and among the various states into which the peninsula had been divided for centuries made unification seem most unlikely. Cavour, who is generally considered the architect of Italian unification, came late to the idea of uniting even northern Italy much less the whole peninsula. That he always wrote in French because his Italian was so bad further illustrates the marvelous character of Italian unification. That the whole of the peninsula was united at all resulted principally from the accident of Giuseppi Garibaldi and his famous Thousand. Cavour had tried vigorously to prevent the irrepressible Genovese from invading Sicily only to have Garibaldi within a matter of months present Cavour's own King Victor Emmanuel of Sardinia with not only Sicily, but the Kingdom of Naples as well. Historian Denis Mack Smith has suggested that the limited energy expended in achieving the Kingdom of Italy is measured in the statistic that more people died in a single day of the Franco-Prussian War than died in all of the twenty-five years of military campaigns to unify Italy.[3] In that story there is little to remind us of the crisis of the American Union.

Can a better analogy be drawn between our war for the Union and the story of German unification? When Otto von Bismarck in 1871 finally brought together into a single nation the heretofore independent states of Germany a new country was thereby brought into existence. The United States, on the other hand, had come into existence almost a century earlier. In 1861 it could hardly be counted as a fledgling state on a par with the newly created German Empire. To make that observation, however, is to read the present back into the past, that is, to assume that the Union of 1787 had created a nation. That, to be sure, is the way Lincoln viewed the Union and, more important, it is the way in which many of us envision the Union, for which a war was necessary in order to excise the cancer of slavery threatening its survival. The unexpressed assumption here is that a *nation* had been endangered, that a sense of true nationhood already embraced the geographical area known as the United States. It was, as noted already, the assumption from which Abraham Lincoln operated. That is why, to respond to a comment made by Robert Bruce, Lincoln, unlike many other American political figures of his time, from whom Professor Bruce quoted, never predicted a war over the Union. A nation does not go to war with itself.[4]

Lincoln's view, however, was not that held by many people of the time, and especially not by Southerners. Suppose we look, then, at the era of the Civil War from the standpoint of the South, and not from the standpoint of him who conquered the region, and denied its essential difference from the rest of the United States. Southerners, it is true, unlike Poles or Hungarians, had originally agreed to join the Union; they were neither conquered nor coerced and they shared a common language, ethnicity, and history. Indeed, the South's sons were among those who drafted the Constitution of the Union, headed the resulting government, and even came to dominate it. Yet, as the early history of the country soon demonstrated, that Union was just a union of states, and not a nation in any organic sense. Paul Nagel in his study of the concept of Union points out that in the first twenty-five years of the country's existence the Union was generally seen as an experiment rather than as an enduring polity. It was, he observes, more a means to achieve nationhood than a nation itself.[5]

Certainly the early history of the country reflects that conception of the Union. Within ten years of the founding of the new government one of the architects of the Revolution and an official of the administration, Thomas Jefferson, boldly asserted a state's right to nullify an oppressive act of Congress. Five years later those who objected to the acquisition of Louisiana talked openly of secession from the Union as a remedy for their discontent, and within another fifteen years even louder suggestions for getting out of the Union came in the course of the war against England. The most striking challenge to the permanence of the Union, of course, came not from New England, but from the South, from South Carolina in particular during the nullification crisis of 1828-33. Just about that time, Alexis de Tocqueville recognized that if the Union was intended to "form one and the same people," few people accepted that view. "The whole structure of the government," he reported, "is artificial," rather than organic.[6] It is true that in 1832 South Carolina stood alone, that not a single state of the South supported its defense of nullification, and some states, like Mississippi, actually branded nullification as naked revolution. But as Tocqueville had implied, that attitude began to change among Southerners as they recognized that their prosperity, racial security, and, in time, their very identity increasingly rested upon slavery.

Historians have disagreed in their conclusions about the growth of Southern nationalism in the course of the thirty years before 1860. Some, like Charles and Mary Beard, saw the Civil War as a conflict between a rising industrial section and a backward agrarian society; others, like David Potter and Kenneth Stampp have played down the differences. Potter went so far as to contend that the "Civil War did far more to produce a

Southern nationalism which flourished in the cult of the Lost Cause than Southern nationalism did to produce the war."[7] Eugene Genovese has argued for a South so different from the rest of the nation as to have a different social system; pre-bourgeois is his word for the South as against a modernizing, bourgeois North. In Genovese's mind slave labor produced a system of values that was peculiar to the South. Genovese has not carried his depiction of a pre-bourgeois South into a discussion of the Civil War or the Confederacy, nor has he and those who agree with him talked about the growth of what other historians have frankly named Southern nationalism.[8] One does not have to go to the lengths that Genovese does to find the emergence of a Southern outlook that could be described as setting the region apart from the rest of the nation. One can find prevalent in the Old South the values of capitalism and modernity, yet still discern there the beginnings of a Southern nationalism.[9]

Nationalism, as historian Drew Faust and others have recently reminded us, is not a commodity or a thing, it is created, brought into being, usually deliberately.[10] It is easy in retrospect to deny the existence of a true sense of national identity among Southerners before 1860, or by 1865, as many historians have done. For we know that once the war was over, that sense of difference among Southerners diminished precipitously—only a handful of Confederates found it necessary to leave the country after Appomattox. At one time historians described nationalism as an organic, almost naturally emerging feeling among a people. In time, it was contended, the feeling or sentiment reached sufficient strength to bring into existence a political framework that united power and feeling in a nation-state. Today, historians are more likely to see nationalism as a process, in the course of which flesh and blood leaders and followers creatively mold and integrate ideas, events, and power to bring a nation into being. That is what happened in the South during the years between nullification in 1832 and Sumter in 1861.

To bring nationalism into being, its proponents need materials to work with, events and personages around which to build and through which to sustain their incipient nationhood. For Southerners the underlying source of that nationalism, of course, was slavery, the wealth-producing capacity of which fixed the South as a region of agriculture and rurality at the very time that the North and West were increasingly diversifying their agriculture with trade, industry, and cities. Slavery, however, was more than a labor system; by the middle years of the nineteenth century, it had become a source of deep political and moral division in the country. It could even be frowned upon by many white Southerners from Jefferson to Henry Clay as contrary to the values of a republic, but its eradication was difficult to accomplish. Increasingly it was seen, by both Southerners and Northerners, as peculiarly Southern.

Once there had been a time, though, when slavery was established in all the states that had fought and won the Revolution. And at one time Americans had worshiped in Baptist, Methodist, and Presbyterian churches that were national in structure. By 1861, however, thanks to slavery, all of these most popular of Protestant denominations had split into Northern and Southern branches. John C. Calhoun himself recognized those old days of common experience. In a letter to his daughter in 1838 he tried to explain to her why secession was not yet the right thing to do. "We must remember," he wrote, "it is the most difficult process in the world to make two people of one. . . . I mean by interior cause of complaint, as in our case, though I do not doubt, if the evil be not arrested in the North, we shall add another example." It did not require nullifiers like Calhoun or secessionists like Robert Barnwell Rhett or William Lowndes Yancey to point out the differences to Southerners, but those Southern peculiarities certainly provided ingredients from which the Rhetts and Yanceys could begin to fashion an ideology of Southern nationalism.

Along with slavery as a source of Southern nationalism went social and economic differences, which, together with the election of an antislavery president, helped to convince many Southerners by 1860 that the Union they had joined in 1787 was not the Union in which they then found themselves. Not only had all the states at the time of the Revolution accepted slavery, but they had all been agricultural in economy, and rural in society as well as proud of their republican ideology that had been fashioned in the course of their joint revolt against Britain's central authority. It did not escape Southerners' attention that the American nationalism being fostered by the expanding urban and industrial economy of the North did not include them or their region. As a Texas politician told the correspondent of the London *Times* in early 1861, "We are an agricultural people. . . . We have no cities—we don't want them. . . . We want no manufactures; we desire no trading, no mechanical or manufacturing classes. . . . As long as we have our rice, our sugar, our tobacco, and our cotton, we can command wealth to purchase all we want."[11] The South's prosperity, which slavery and the plantation had generated, only deepened the divisions between the regions and sharpened the recognition that Northerners were not like Southerners, that the South was a different place, that Southerners were strangers in the house of their fathers. As historian Allan Nevins later wrote, "South and North by 1857 were rapidly becoming separate peoples. With every passing year, the fundamental assumptions, tastes, and cultural aims of the sections became more divergent." The North, wrote Southern novelist William Gilmore Simms in 1852, presses the South on the slavery question.

"But we are *a nation,*" he replied, "with arms in our hands, and in sufficient numbers to compel the respect of *other nations.* . . ."[12]

This recognition of the Union's transformation since its founding is plainly reflected in Confederates' frequent insistence that their cause was but a rerun of the Revolution against England. It is surely no coincidence that February 22 and July 4 were official holidays of the Confederacy or that Jefferson Davis was inaugurated on Washington's birthday. In both 1776 and 1860, objections to political impositions by the dominant power were prominent, but an equally powerful source of the two revolutions was a sense of being alien, of being an outsider, a perception that independence would remove.

The existence of Southern nationalism, even the attenuated variety that I am asserting here, is admittedly not a settled issue among historians. Indeed, I suspect that most historians agree with Kenneth Stampp and think of its assertion in the late 1850s as a subterfuge, almost a trick played upon the mass of Southerners by a relatively few so-called Southern "fireaters" like Yancey and Rhett. And certainly there were many men and women in the South in 1860 who spoke of themselves as Unionists. Lincoln, too, along with many other Republicans, thought Southern alienation from the North was but a ploy to gain concessions. Yet, despite such widely held doubts, in 1861 eleven Southern states withdrew from the Union and then proceeded to fight the bloodiest war of the nineteenth century to defend that decision. The proportion of Southerners who died in that struggle far exceeded that experienced by Americans in any other war and was exceeded during the Second World War only by the losses sustained by Germans and Russians. That straightforward quantitative fact, I think, provides the most compelling response to Kenneth Stampp's doubts that Southerners were committed to winning.[13]

We call the struggle the Civil War, some Southerners who accepted the Southern view of the Constitution, call it the War Between the States, and officially it is the War of the Rebellion. But it was, of course, really the War for Southern Independence, in much the same league, if for different historical reasons, as Poland's and Hungary's wars of national liberation around the same time. We know, too, that the South's determined struggle revealed how wrong Lincoln had been to believe in a broad and deep sense of Unionism among Southerners.

European observers of the time well recognized the incomplete nature of American nationalism, if Lincoln did not. William Gladstone, the English Chancellor of the Exchequer in 1862, could not conceal his conviction, as he phrased it, that "Jefferson Davis and other leaders of the South have made an army; they are making, it appears, a navy; and they have made what is more than either, they have made a nation." Soon after the war the great liberal historian Lord Acton, in a letter to Robert E. Lee, explained why he had welcomed the Confederacy. "I saw in State Rights," Acton wrote, "the only availing check upon the absolutism of the sovereign will, and secession filled me with hope, not as the destruction but as the redemption of Democracy. . . . I deemed that you were fighting the battles of our liberty, our progress, and our civilization; and I mourn the stake which was lost at Richmond more deeply than I rejoice over that which was saved at Waterloo." [14]

In short, when the South seceded in 1860–61 that fact measured not only the failure of the Union, but, more important, the incomplete character of American nationalism. Or as historian Erich Angermann has reminded us, the United States in 1861, despite the Union of 1787, was still an "unfinished nation" in much the same way as were Italy and Germany. [15]

True, a deep sense of nationhood existed among Ameri-

cans, but it was confined largely to the North. Indeed, to acknowledge that nationalism is probably the soundest way to
account for the remarkable explosion of popular support that
greeted Lincoln's call for volunteers to enforce the laws in the
South after the fall of Sumter. When we recognize that in 1860
only a truncated nationalism existed among Americans despite
the eighty-year history of the Union, then the American Civil
War suddenly fits well into a comparison with other nation-
building efforts of those years. The Civil War, in short, was
not a struggle to save a failed Union, but to create a nation
that until then had not come into being. For, in Hegel's ele-
gant phrase "the owl of Minerva flies at dusk," historical understanding is fullest at the moment of death. International
comparison throws into relief the creative character of war in
the making of nations, or, in the case of the Confederacy, in
the aborting of nations.

For one thing, all of the struggles for national unification
in Europe, as in the United States, required military power to
bring the nation into existence and to arm it with state power.
This was true not only of Italy and Germany, but of Switzerland as well, as I hope to show a little later. As Ernest Renan
wrote in his 1882 essay "What Is a Nation?," "Unity is always
realized by brute force. The union of North and South in
France," he pointed out, "was the result of a reign of terror
and extermination carried on for nearly a century" in the late
Middle Ages. "Deeds of violence . . . have marked the origin
of all political formations," he insisted, "even of those which
have been followed by the most beneficial results."[16]

The Italian wars of national unity may not present much of
an analogy with the American war, but the course of German
unification is revealing. Everyone is familiar with the role of
the Franco-Prussian war in the achievement of the unification

of Germany in 1871. Equally relevant for an appreciation of the American Civil War as struggle for nationhood was the Seven Weeks War between Austria and Prussia, which preceded the war with France and which culminated in Prussia's great military victory at Königgrätz or Sadowa in 1866. That war marked the culmination of Bismarck's determined efforts to exclude Austria from any united Germany in order that Prussia would be both the center and the head. By defeating Austria and creating the North German Confederation under the leadership of Prussia, Bismarck concluded what many observers at the time and historians since have referred to as a *Bruderkrieg*, a German civil war.[17] For it was neither foreordained by history nor by the power relations among the states of central Europe that a *Kleindeutschland* or lesser Germany from which Austria was excluded would prevail over a *Grossdeutschland* or greater Germany in which Austria would be the equal or even the superior of Prussia.

At that stage in the evolution of German nationhood, the closest analogy to the American experience puts Prussia in the position of the Southern Confederacy, for it was in effect seeking to secede from the German Confederation, created at the time of the Congress of Vienna and headed by Austria. Just as Bismarck had provoked Austria into war to achieve his end, so Jefferson Davis and the South were prepared to wage war against their long-time rival for control of the North American Union.

Despite the tempting analogy, however, Jefferson Davis was no Bismarck. His excessive constitutional scruples during the short life of the Confederacy make that crystal clear. (If anything Bismarck was just the opposite: slippery in regard to any constitution with which he came into contact.[18]) Davis's rival for domination of the North American continent—Abraham Lincoln—came considerably closer to Bismarck, including the

Two Men of Blood and Iron? ABOVE: Bismarck in 1862 and RIGHT:
Lincoln in 1864. Photographs by anonymous and Mathew Brady.
Courtesy of the Lincoln Museum, Fort Wayne, Indiana.

Lincoln.

Bismarck who by his innovative actions within the North German Confederation had laid the foundations of German industrialization.[19]

Historians of the United States have not liked comparing Bismarck and Lincoln. As historian David Potter once wrote, "the Gettysburg Address would have been as foreign to Bismarck as a policy of 'blood and iron' would have been to Lincoln."[20] It is certainly true that the Gettysburg Address could not have been a policy statement from Bismarck, though he boldly introduced universal manhood suffrage and the secret ballot in the new Germany, much to the horror of his conservative friends and to the consternation of his liberal opponents. And it is equally true that the Junker aristocratic heritage and outlook of the mature Bismarck stands in sharp contrast to the simple origins and democratic beliefs of Abraham Lincoln. But if we return to seeing the war and Lincoln's actions at the time from the standpoint of the South then the similarities become clearer. Once we recognize the South's disenchantment with the transformation in the Union of its fathers and its incipient nationalism, which slavery had sparked, we gain an appreciation of the incomplete nature of American nationalism. Lincoln then emerges as the true creator of American nationalism, rather than as the mere savior of the Union.

Given the immense carnage of the Civil War, not to mention the widespread use of iron in ordnance and railroads, that struggle in behalf of American nationality can hardly escape being described literally as the result of a policy of blood and iron. The phrase fits metaphorically almost as well. Reflect on Lincoln's willingness to risk war in 1861 rather than compromise over the issue of slavery in the territories. "The tug has to come, and better now, than anytime hereafter," he advised his fellow Republicans when the Crittenden compromise was before Congress.[21] Like Horace Greeley, Lincoln was determined to call what he considered the South's bluff, its frequent threat over the years to secede in order to extract one more concession to ensure the endurance of slavery. Convinced of the successful achievement of American nationhood, he counted on the mass of Southerners to rally around the national identity, only to find that it was largely absent in the region of his birth. Only military power kept even his native state within the confines of his nation. Bismarck had to employ no such massive power to bring the states of south Germany into his new Reich in 1870–71. Rather, their sense of a unified Germany bred over a quarter-century of common action brought Catholic Bavaria, Württemberg, and Baden immediately to Protestant Prussia's side when France declared war in 1870.

But then, unlike Bismarck, Lincoln was seeking to bring into being a nation that had lost whatever sense of cohesion its Union of 1787 may have nurtured. His task was more demanding and the means needed to achieve the goal were, for that reason, harsher, more deadly, and more persistently pressed than the creation of a new Germany demanded of Bismarck. Lincoln's commitment to nationhood rather than simply to the Union comes through quite clearly in an observation by James McPherson.[22] In his First Inaugural, Lincoln used the word "Union" twenty times; "nation" appears not at all. (That description of the United States, of course, had long been anathema to the South.) Once the South had seceded, however, the dread word began to appear in his texts: three times in his first message to Congress. By the time of the Gettysburg Address, the term "Union" appeared not at all, while "nation" was mentioned five times. In his Second Inaugural, Lincoln used Union only to describe the South's actions in disrupting the Union in 1861; he described the war as having saved the "Nation," not simply the Union.

In deeds as well as in words, Lincoln came closer than Jefferson Davis to Bismarck. There is nothing in Lincoln's record that is comparable to Bismarck's famous "Ems dispatch" in which he deliberately edited a report on the Prussian king's reaction to a demand from the French government in such a way as to provoke the French declaration of war that Bismarck needed in order to bring the south German states into his unified Germany. Over the years, the dispute among United States historians whether Lincoln maneuvered the South into firing the first shot of the Civil War, has not reached the negative interpretation that clings to Bismarck's Ems dispatch. Yet Lincoln's delay in settling the issue of Sumter undoubtedly exerted great pressure upon the Confederates to fire first. To that extent his actions display some of the earmarks of Bismarck's maneuvering in 1870. For at the same time Lincoln was holding off from supplying Sumter he was firmly rejecting the advice of his chief military adviser, Winfield Scott, that surrendering the fort was better than provoking the Confederates into beginning a war. Lincoln's nationalism needed a war, but one that the other side would begin.[23]

The way in which Lincoln fought the war also reminds us at times of Bismarck's willingness to use iron, as well as shed blood, in order to build a nation. Throughout the war Lincoln denied that secession was a legal remedy for the South, yet his own adherence to constitutional limits was hardly flawless. If Bismarck in 1862 in behalf of his king's prerogative interpreted parliamentary government out of existence in Prussia for four years, Lincoln's interpretation of the American Constitution followed a similar, if somewhat less drastic path. As Lincoln scholar James G. Randall remarked years ago, Lincoln employed "more arbitrary power than perhaps any other President. . . . Probably no President has carried the power of proclamation and executive order (independently of Congress) as far as did Lincoln." Randall then proceeded to list those uses of power: freeing slaves, accepting the dismemberment of Virginia by dubious constitutional means, providing for the reconstruction of states lately in rebellion, suspending the writ of habeas corpus, proclaiming martial law, and enlarging the army and the navy and spending public money without the necessary Congressional approval. "Some of his important measures," Randall points out, "were taken under the consciousness that they belonged within the domain of Congress. The national legislature was merely permitted," Randall continues, "to ratify his measures, or else to adopt the futile alternative of refusing consent to accomplished fact."[24] Lincoln himself justified his Emancipation Proclamation on the quite questionable ground "that measures otherwise unconstitutional might become lawful by becoming indispensable to the preservation of the Constitution through the preservation of the nation."[25]

That slavery was the spring and the river from which Southern nationalism flowed virtually dictated in Lincoln's mind that it must be extirpated for nationalist as well as humanitarian reasons. For many other Northern nationalists the fundamental role slavery had played in the creation of Southern nationalism must have been a prime reason for accepting its eradication. Few of them, after all, had been enemies of slavery in the South, much less friends of black people. Indeed, hostility to blacks on grounds of race in the 1860s was almost as prevalent in the North as in the South.

What the war represented, in the end, was the forceful incorporation of a recalcitrant South into a newly created nation. Indeed, that was exactly what abolitionist Wendell Phillips had feared at the outset. "A Union," he remarked in a public address in New York in 1860, "is made up of willing States. . . . A husband or wife who can only keep the other partner within the bond by locking the doors and standing armed before them,

Assassination. ABOVE: Currier & Ives lithograph of the Lincoln assassination from 1865. BELOW: Woodcut from 1866 depicting an attempt on Bismarck's life. *Courtesy of the Lincoln Library, Fort Wayne, Indiana.*

had better submit to peaceable separation." The United States, he continued, is not like other countries. "Homogeneous nations like France tend to centralization; confederations like ours tend inevitably to dismemberment."[26]

A similar objection to union by force had been advanced by none other than that old nationalist John Quincy Adams. "If the day should ever come (may Heaven avert it)," he told an audience celebrating the jubilee of the Constitution in 1839, "when the affections of the people of these states shall be alienated from each other; when the fraternal spirit shall give away to cold indifference . . . far better will it be for the people of the disunited states, to part in friendship from each other, than to be held together by constraint."[27] In Lincoln's mind, it was to be a stronger and more forceful nation, one which would mark a new era in the history of American nationality, just as Bismarck's proclamation of the new German Empire in the Hall of Mirrors at Versailles in January, 1871, constituted both the achievement of German unity and the opening of a new chapter in the history of German nationality.

The meaning of the new American nationhood as far as the South was concerned was its transformation, the rooting out of those elements that had set it apart from Northern nationalism. In the context of nation-building the era of Reconstruction can best be seen as the eradication of those aspects of the South that had lain at the root of the region's challenge to the creation of a nation. That meant ridding the South not only of slavery, but also of its undemocratic politics, its conservative social practices, its excessive dependence upon agriculture, and any other habits that might prevent the region from being as modern and progressive as the North.

Nowhere does this new nationalism appear in more strident form than in an essay by Senator Charles Sumner deceptively entitled "Are We a Nation?"[28] The title was deceptive because there was no doubt in Sumner's mind that the United States was indeed a nation, and had always been. The essay was first given as a lecture in New York on the fourth anniversary of Lincoln's delivery of the address at Gettysburg. Sumner was pleased to recall Lincoln's reference to "a new nation" on that previous occasion, causing Sumner to remark that "if among us in the earlier day there was no occasion for the Nation, there is now. A Nation is born," he proudly proclaimed.[29] That new nation, he contended, was one in behalf of human rights, by which he meant the rights of blacks, which the South must now accept and protect.

Interestingly enough, in the course of his discussion of nationhood, Sumner instanced Germany as a place where nationhood had not yet been achieved. "God grant that the day may soon dawn when all Germany shall be one," he exclaimed.[30] In 1867 he could not know what we know today: that the defeat of Austria at Königgrätz the year before had already fashioned the character and future of German unity under Bismarck.

No single European effort at creating a new sense of nationhood comes as close to that of the United States as Switzerland's. Although the Swiss Confederation, which came into existence at the end of the Napoleonic era, lacked some of the nationalist elements of the American Constitution, it constituted, like the United States, a union of small states called cantons, which, again like the states of the American Union, had once been independent entities. And as was the case in the American Union, the cantons of the Swiss Confederation were separated by more than mountainous terrain. The role that slavery played in dividing the United States was filled among the Swiss by religion. The Catholic cantons of Uri, Schwyz, and Unterwalden had been the original founders of the confederation in the days of William Tell, while the Protestant cantons were not only the more recent, but more important, the cantons in which the liberal economic and social ideas and forces that were then reshaping European society had made the most headway.

Among the intellectual consequences of that modernity was a growing secularism, which expressed itself in 1841 in the suppression of all religious orders by the Protestant canton of Aargau. The action was a clear violation of the Federal Pact of 1815, but none of the Protestant cantons objected to it. The Catholic cantons, however, led by Lucerne, vehemently protested the overriding of their ancient rights. In this objection there is a striking parallel with the South's protest against the North's attacks on slavery and refusal to uphold the fugitive slave law; both slavery and a fugitive slave law, of course, were embedded in the original United States Constitution.[31]

The Catholic cantons' response to the violation of the Confederation's constitution was that Lucerne then invited the Jesuit Order to run its schools, much to the distaste of the Protestants in Lucerne and the Protestant cantons in general. Some of the Lucerne liberals then set about to organize armed vigilantes or *Freischaren* to overthrow the governments in the Catholic cantons. The American analogy for these military actions that leaps to mind, of course, is "bloody Kansas." Nor was the guerrilla violence in Switzerland any less deadly than that in Kansas. When the canton government of Lucerne sentenced a captured *Freischar* to death, a group of his supporters invaded the canton and triumphantly carried him off to Protestant Zürich. More than one hundred died in the escapade.

Like "Bloody Kansas," the guerrilla phase of the Swiss conflict between old (Catholic) and new (Protestant) cantons deepened a sense of alienation between the two contending parties, which, in turn, led, almost naturally, to a move for separation from the Confederation. In December, 1845, seven Catholic cantons, including, interestingly enough, the three founding cantons of ancient Switzerland, formed what came to be called the *Sonderbund* or separatist confederation. Unlike the Southern states in 1860–61, the cantons of the *Sonderbund* did not proclaim secession, though they clearly saw themselves as resisting violations of traditional constitutional rights. Indeed, under the rules of the Swiss Confederation regional agreements among cantons were permissible, but the army the *Sonderbund* cantons brought into being, and the public stands they announced, strongly suggested to the rest of Switzerland that secession was indeed their intention. And so in July, 1847, the Diet of the Confederation ordered the *Sonderbund* to dissolve, an act that precipitated the departure of the delegates of the *Sonderbund* cantons. Again, like the Confederacy, the *Sonderbund* sought foreign support (particularly from Catholic and conservative Austria), but it was no more successful in that respect than the Confederacy. In early November the Diet voted to use force against the *Sonderbund;* civil war was the result.

Although each side mustered 30,000 or more troops under its command, the war was brief and light in cost; it lasted no more than three weeks and fewer than 130 men lost their lives. The victory of the Confederation's forces resulted in the rewriting of the constitutional relations among the cantons. The new national government was to be a truly federal republic deliberately modeled after that set forth in the Constitution of the United States.[32] The immediate postwar era in Switzerland exhibited little of the conflict that we associate with the Reconstruction era. But then, the Swiss civil war was short and if not sweet, at least not very bloody. Yet there, too, as in the United States, the winners deemed it essential to extirpate those institutions that had been at the root of the disruption of the Confederation. Before the cantons of the *Sonderbund* were accepted back into the Confederation they were compelled to accede to barring the Jesuit Order from all the cantons. The acceptance of the Order into Lucerne had been, after all, a major source of the cantonal conflicts that led to the civil war.[33] A measure of the depth of the religious issue in the Swiss conflict is that almost a century and a half passed before the Jesuit Order was readmitted to Switzerland. And in that context it is perhaps worth remembering that a century passed before a president of the United States—Lyndon B. Johnson—could be elected from a state of the former Confederacy.

As happened with the Civil War in the United States, the *Sonderbundkrieg*—the war of the Separatist Confederation— marked the long-term achievement of nationhood. So settled now was the matter of Swiss national identity that when Europe erupted in 1848 in wars of national liberation and revolution, the new Swiss Federation, the embodiment of Swiss nationality, escaped entirely from the upheaval. No longer was there any question that Switzerland was a nation; just as after 1865 there could be no doubt that the United States was a nation. In both instances, war had settled the matter for good.

Finally, there remains yet one more comparison between America's achievement of nationhood through war and the unification of Germany. Contemporaries in 1871 and historians since often saw in the creation of the *Kaiserreich* something less than a comforting transformation of the European international scene. It is true that the new Empire did not include all European Germans within its confines. That is why Bismarck was seen as a *Kleindeutscher*. But never before had so many Germans been gathered within a single state and especially one with a highly trained and efficient army, as the quick defeat of Austria in 1866 and of France in 1870 forced everyone to recognize. The military presence of Prussia under Frederick the Great, once so formidable in central Europe, was easily surpassed by the new empire of his Hohenzollern descendants. It was an empire whose power would soon challenge its neighbors and the peace of Europe, despite Bismarck's original aim of hegemony without more war.

If nationhood through the agency of war meant a Germany of new power and potential danger to others, the achievement of nationhood by the United States during its civil war carried with it some strikingly similar aspects. Out of the war, the United States emerged, not only a nation, but also by far the strongest military force in the world of the time. But with the United States, as with the new German empire, military might was not the only source of a new tone in relations with other states. Nationhood brought a new self-confidence, even self-assertion, that ignited the apprehensions of neighbors. Even before its mighty victory over the South, the United States had been perceived in Europe as a rambunctious, even irresponsible Republic, challenging when not overtly rejecting the traditional ways of Europe and of international relations. As a European power, the new German empire aroused the fears of Europeans as the enhanced power of the United States, being separated from Europe by the Atlantic, never could. But those European powers which had interests in the New World soon found that the enlarged authority of the United States could well spell danger.

The first to sense it were the French, who had presumed to meddle in the internal affairs of Mexico while the United States had been preoccupied with suppressing the division within its own borders. The defeat of the Confederacy allowed the triumphant United States to turn upon the French for threatening American hegemony in the New World, a threat that never needed to be implemented since the Mexican forces themselves soon routed the meddling French. As far as the southern neighbor was concerned, the achievement of nationhood by the United States could be seen, temporarily, at least, as supportive rather than threatening.

For the neighbor to the north, the story was rather different. Ever since their founding revolution in 1776, Americans have thought that the most natural thing in the world would be for the English-speaking people to the north to join the United States. Though most Canadians, then as now, have rejected annexation, some Canadians have always thought it was natural and inevitable. The threat of annexation reached a new height during the Civil War, especially after some Confederate agents managed to mount a successful military raid from Canada against St. Albans, Vermont. The outrage expressed by the government in Washington, coupled with new talk of annexation aroused both Canadian nationalists and British statesmen to seek ways to counter Canada's vulnerability to the power of the newly emboldened American nation.

In the age of the American Civil War, the country known today as Canada, was a collection of diverse governmental and even private units, some of which were self-governing, but all of which were parts of the British Empire. The move to create a united Canada was spawned not only by a fear of annexation by the United States, but by an even more compelling insight from the American Civil War. It was the lesson that a vaguely defined federal system such as that of the United States could end up in civil war. The upshot was that the federal constitution drawn up in 1867 (technically known as the North America Act) to unite all of Canada under one government, placed all residual powers in the hands of the national government, a lesson derived from the perceived result in the United States of leaving to the states those powers not specifically granted to the federal government. As historian Robin Winks has remarked, "In effect the war had helped create not one but two nations."[34]

But was the Canadian union established in 1867 a nation? Winks uses that term, but is it a nation in the organic sense that we have been talking about here? Listen to his own summary description of Canadian nationhood: "Born in fear, deadlock, and confusion, Canada grew into a nation that could not afford to exhibit the rampant nationalism usually associated with young countries, and even today [that is, in 1960], due to her mother, the nature of gestation, and the continuing pressures from her large, pragmatic and restless neighbor, Canada remains a nation in search of a national culture."[35]

Canada, of course, was one of these countries that between 1845 and 1870 struggled to achieve a truly national identity. Does the Canadian example offer any further insight into the meaning of that era of nation-building in general or of Lincoln's nation-building in particular? The Canadian experience, I think, puts the cap on the argument I have been making throughout these remarks. And that brings us back to the present, with which I began. Let me reach my conclusion with a personal anecdote.

Not so many years ago I asked several German historians of my acquaintance whether they thought the division of Germany that the Cold War had caused would ever be healed. The general response was that there was little reason to believe that after forty years of division the two Germanies, so dissimilar now in economy, politics, and culture, would have much in common. Most Germans, they added, were too young to have ever even experienced a united Germany. But what about the united Germany that Bismarck had created, I asked out of my deeply held belief in the power of history. Oh, they responded, you forget that a united Germany has a short existence in the long history of Germans: a mere seventy years, after all, from 1871 to 1945.[36]

Today, of course, we know that those seventy years were controlling, that the Germany of Bismarck has endured despite a period of division that is more than half the length of its

years of unification. Such is indeed the power of history. Contrast that picture from today's world with that of Canada. The
Canadian union of 1867 is still jeopardized by ethnic and other
differences, despite the efforts by some leading Quebeckers to
smooth over the divergences between French-speaking and
English-speaking Canadians. How then does the German or the
Swiss, or the American road to nationhood differ from that of
the Canadians? Obviously there are a number of cultural and
historical differences, but one that grabs our attention in this
comparison is that only Canada failed to experience a war of
national unification. During the nullification crisis in 1832 John
Quincy Adams remarked to Henry Clay that "It is the odious
nature of the [Union] that it can be settled only at the cannon's mouth."[37] But as Lincoln recognized and Ernest Renan
reminded us, it was a nation, not merely a Union, that blood
and iron brought into existence.

5

One Alone?
The United States and
National Self-determination

KENNETH M. STAMPP

*I*N THE AUTUMN OF 1991 we were stunned by the political collapse of one of the world's so-called superpowers, the resulting success of the Baltic republics in regaining their independence, and the disintegration of the rest of the Soviet Union into a loose economic federation of autonomous or independent republics. Elsewhere in Europe, notably in Czechoslovakia and Yugoslavia, religious and other ethnic differences seriously threaten national unity. Few modern national states, in fact, are without substantial ethnic minorities, and it is always tragic when a government finds no better way to avoid political fragmentation than by the use of troops and tanks. The alternative is to accommodate minorities either by respecting their ethnic identities or by granting them a measure of autonomy within the framework of a less centralized nation. In some cases, when a substantial majority of a localized disaffected ethnic group desires independence, a negotiated separation would seem preferable to military control. A nation held together with bayonets alone is likely to be terminally ill.

The United States has a long tradition—not unmixed, of course, with self-interest—of sympathy for movements abroad which sought to vindicate the doctrine of national self-determination—if I may use a twentieth-century term for a nineteenth-century nationalist concept. Early in the nineteenth century North Americans greeted the Latin American revolutions for independence from Spain and Portugal as replications of their own struggle for self-government. "They adopt our principles," boasted Henry Clay, "copy our institutions, and, in many instances, employ the very language of our revolutionary papers." The Monroe Doctrine of 1823, in part a defense of self-determination, warned European powers not to violate the sovereignty of the independent nations of the New World.[1] During the revolutions of 1848, politicians such as Daniel Webster, William H. Seward, and Stephen A. Douglas expressed enthusiastic support for Hungary's struggle to free itself from Austrian rule. After Austria, aided by Russia, had crushed the rebellion, the American public lionized its exiled leader, Louis Kossuth, during an extended tour of the United States.[2] In the 1860s and 1870s the unification of Germany and Italy, as expressions of self-determination, elicited a favorable American response. Cuba ultimately won its independence from Spain with American military assistance, although motives in this instance were a good deal more complex.

National self-determination was almost formalized as an American doctrine during the first World War when Woodrow Wilson, in an address to a joint session of Congress, January 8, 1918, announced his famous Fourteen Points. Hoping to make them the foundation of a just and lasting peace, he devoted eight of his points to specific applications of the principle of self-determination—among them, the return of Alsace-Lorraine to France, the restoration of an independent Poland, and the provision of opportunities for "autonomous development" of ethnic groups within the Austro-Hungarian and Ottoman empires.[3]

There have obviously been occasions when the intrusion of economic or strategic considerations has compromised our traditional respect for the right of self-determination. The Spanish-American War freed the Philippines from Spanish rule, but freedom from American rule was long delayed. Wilson himself was a party to several violations in the Treaty of Versailles. Our government felt obliged, however reluctantly, to tolerate some egregious violations after the second World War. Moreover, throughout the twentieth century, even after the formulation of President Franklin Roosevelt's Good Neighbor policy, America has frequently meddled, both overtly and covertly, in the internal affairs of the republics in Central America and the Caribbean. The intervention has been so persistent as to suggest a reluctance among our policy-makers to permit self-determination to operate in our own backyard. These violations of a long-standing tradition have seldom gone unchallenged. They have often provoked substantial and, occasionally, effective protests in Congress, in the press, and from various civic and ad hoc organizations.

However, I wonder what the public response would be if the question of self-determination should arise as an American *internal* issue, as it now presents itself in many other nations, including our Canadian neighbor. Canada confronts the problem in French-speaking Quebec, whose gradually escalating demands for autonomy may yet culminate in a movement for full independence. The Canadian government has elected to deal with the problem peacefully through negotiation, and a resort to a violent resolution at present appears unlikely.

The idea that such a problem might one day confront the United States government might seem preposterous, but since the future is unpredictable a little speculation may be justified. Suppose the large Spanish-speaking population in Florida, or southern Texas, or southern California continues to grow until it becomes in one or another of these regions a governing majority. Suppose also that within this population ethnic consciousness increases and grievances and resentments accumulate, culminating, as in Quebec, in a movement for political autonomy. Would the tradition of self-determination prevail, or would a second tradition—the one sealed at Appomattox after the loss of 600,000 American lives—be invoked? Abraham Lincoln defined the second tradition succinctly in his first inaugural address: "I hold, that in contemplation of universal law, and of the Constitution, the Union of these States is perpetual. . . . Continue to execute all the express provisions of our national Constitution, and the Union will endure forever—it being impossible to destroy it, except by some action not provided for in the instrument itself."[4] In short, as the South learned at a terrible price, self-determination was not applicable to the federal Union. Apparently there was one tradition for export, another for use at home.

The second tradition, which Lincoln upheld with such relentless determination, had evolved only slowly and uncertainly over many years following the American Revolution.[5] Although the Articles of Confederation had explicitly stated that "the Union shall be perpetual," the Constitution did not settle that question with such clarity. Nationalists could only infer the Union's perpetuity from certain passages in the document, all of which were subject to more than one interpretation. Even Lincoln seemed to concede that the language of the Constitution was not conclusive. Taking another tack, he argued that perpetuity "is implied, if not expressed, in the fundamental law of all national governments. It is safe to assert that no government proper, ever had a provision in its organic law for its own termination."[6]

Neither the debates in the Philadelphia convention nor those in the various state ratifying conventions revealed a general understanding that the Union formed by the Constitution was to be perpetual. James Madison, in one of his contributions to *The Federalist,* assured the public that the Constitution was to be ratified "by the people, not as individuals composing one entire nation; but as composing the distinct and independent States to which they respectively belong." Although Madison later recalled that it had been the "sincere and unanimous wish" of the Philadelphia delegates that the Union would be preserved, several of them were quite philosophical about the possibility that it would eventually dissolve.[7]

For many years after the new Constitution had been ratified the general consensus was that the Union was an experiment, valued not as a good in itself but as a device to achieve certain valuable ends. The Union, John Randolph asserted in 1814, was "the *means* of securing the safety, liberty, and welfare of the confederacy and not itself an end to which these should be sacrificed." There was at that time little evidence of an American national identity that would encourage a belief in the Union as an absolute good.[8] Fear that the Union would divide along sectional lines arose as early as the 1790s, and President Washington alluded to it in his Farewell Address. "Is there doubt whether a common government can embrace so large a sphere?" he asked. He was not sure of the answer but urged that it be left to experience, for it was "well worth a fair and full experiment."[9] Indeed, as Robert Bruce had shown, the fear of a violent civil war also remained alive.

A few years later, Madison and Jefferson wrote a series of resolutions, adopted by the Virginia and Kentucky legislatures, formulating a theory of the Union that eventually would be used to justify southern secession. The Constitution, they contended, was a compact agreed to by the sovereign states, each of which retained the right to "interpose" its authority against unconstitutional federal acts and to decide upon the appropriate remedy. During his presidency, Jefferson, who never questioned the right of a state to secede, speculated about a possible separation of the eastern and western states at some future time. "God bless them both," he wrote, "and keep them in the union if it be for their good, but separate them if it be better."[10] Meanwhile, during the Jefferson and Madison administrations, the disaffection of New England Federalists grew so intense that some began to doubt the value of the Union. Timothy Pickering of Massachusetts found "no magic in the sound of Union. If the great objects of union are utterly abandoned . . . let the Union be severed. Such a severance presents no Terrors for me."[11]

As late as the 1820s the general view of the Union as an experiment still persisted, and the alleged right of secession had not yet been challenged by a systematic argument affirming the perpetuity of the Union. Nevertheless, the state of the Union had by then changed significantly. Its practical economic value was widely understood. Moreover, during and after the War of 1812, a strong sense of nationhood and pride in American citizenship had developed. The United States did not escape the current of romantic nationalism that was sweeping over the Western world. The time was ripe for a new conception of the federal Union as an absolute good.[12]

Massive support for the idea of an unbreakable Union first developed in the early 1830s when South Carolinians attempted to nullify the federal tariff laws and threatened secession. The nationalists responded with the first elaborate arguments for perpetuity. Among them the most comprehensive was Presi-

dent Andrew Jackson's Proclamation on Nullification of December 10, 1832, prepared for him by Secretary of State Edward
Livingston. Neither Lincoln, in his first inaugural address, nor
the Supreme Court, when it finally took up the matter in *Texas
v. White* in 1869, added much to Jackson's case. The Constitution, he asserted, was no mere compact between sovereign states.
Rather, it had been framed and ratified by the people. They
had formed "a *government,* not a league," and it operates directly on the people, not the states. The Supreme Court, rather
than the states, is the proper authority to settle controversies
arising under the Constitution. Secession is not a constitutional remedy reserved to the states but an act of revolution,
and the duty of the President is, according to his oath, "to take
care that the laws be faithfully executed."[13] Jackson's argument
had its flaws, both historical and logical, but so did the argument supporting the right of secession. "It is the odious nature
of the question," John Quincy Adams once observed, "that it
can be settled only at the cannon's mouth."[14]

In the years that followed, the increasingly disaffected South
remained the last stronghold of the old and once widely respected concept of the Constitution as a compact and of the
Union as a voluntary federation of sovereign states. Yet, hardly
anyone, including the nationalists for whom the Union had
become an absolute good, wanted the question to be settled
by force. Even Jackson had hoped to avoid bloodshed.[15] As
sectional tensions mounted, most conservative Unionists tried
desperately to avoid political disruption by urging compromises rather than a violent confrontation. To substitute military coercion for the voluntary loyalty of the past, many feared,
would deform the Union and transform the federal government into a Leviathan under which civil liberties would perish.
By the 1850s a more militant Unionism was emerging among
antislavery Republicans, many of whom seemed ready to meet

the southern secessionists head-on.[16] But conservative northern and border-state Democrats and old Whigs, although firm Unionists, still viewed such a collision with dread.

Finally, after Lincoln's election, southern secession, which had been threatened so often in the past, became a reality. In justifying their action, some political leaders in the eleven states that eventually seceded invoked the American revolutionary tradition and identified themselves with the patriots of 1776. More commonly they claimed to be resorting to a constitutional right reserved by the states as a remedy for intolerable violations of the federal compact. South Carolina, in its "Declaration of the Causes of Secession," affirmed its sovereignty, enumerated its grievances (all involving northern attacks on slavery), and announced that it had "resumed [its] position among the nations of the world, as a separate and independent State."[17]

Although the term "self-determination" was not then in use, southern secession was in essence an assertion of that right. But this movement had its oddities. First, the three million enslaved blacks, the Confederacy's true ethnic minority and its most severely oppressed population, had no voice in the matter and no reason to support secession. Second, white Confederates, in spite of their claim that they had become a distinct and separate people, had no ethnic characteristics to distinguish them from Northerners—no notable differences in language, religion, political traditions, or population origins, and few unique traits to give them a clear cultural identity.[18] Rather, at the core of the white South's drive for independence was its perception of Lincoln as a threat to its slave labor system and its conviction, based on racial fears, that emancipation would be an economic and social catastrophe. The "immediate cause" of secession, according to Alexander H. Stephens, Vice President of the Confederacy, was slavery. In the Confederate government, he avowed, "its foundations are laid, its cornerstone rests, upon the great truth that the negro is not equal to the white man; that slavery . . . is his natural and moral condition."[19]

When northern nationalists were at last convinced that the disunionists were in earnest, they ridiculed the constitutional justification of peaceful secession as a feeble argument long since discredited. Southern ordinances of secession, in Lincoln's blunt words, were "insurrectionary or revolutionary, according to circumstances."[20] As for the *right* of revolution, proclaimed by Thomas Jefferson, that drastic course could be justified only in a struggle *against* oppression, not in its defense. The Philadelphia *Press* doubted that any revolution was ever commenced "on more trifling and trivial grounds" than those advanced by the South.[21]

Yet, standing on the brink of a great national crisis, many Unionists faltered briefly before making their fateful decision to deny, with whatever force was necessary, the right of self-determination to their dissatisfied southern countrymen. At the last minute a few Republicans toyed with the idea of letting them, after all, depart in peace, as the abolitionist followers of William Lloyd Garrison had urged for many years.[22] More frequently those who hoped to avoid violence found an attractive alternative in what they described as a "Fabian policy" of "masterly inactivity." All that was required of Northerners, they said, was to remain calm, avoid threats, and wait patiently for Southerners voluntarily to return to their allegiance.[23] Senator Seward of New York, who would become Lincoln's Secretary of State, was a firm exponent of this policy and did not abandon it until Confederate guns opened fire on Fort Sumter.[24]

Meanwhile, throughout the secession winter, northern Democrats and border-state Unionists turned to the traditional Union-saving remedy: compromise. Congressional Republicans, however, blocked every proposed compromise that involved a retreat from the principles of their national party platform. Moreover, by February 1861 the Senators and Representatives from the Deep South, having themselves shown little interest in compromise, had resigned and gone home. Although the tradition of sectional compromise dated back to the Constitutional Convention of 1787, those who tried to uphold that tradition failed for two obvious reasons: first, because southern secessionists insisted that the time for compromise had passed and, second, because most northern Republicans, on that point at least, agreed with them.[25]

From the very beginning of the crisis a powerful contingent of "stiff-backed" Republicans had openly asserted that, sooner or later, the secessionists would have to be suppressed by force and that there would be no better time for that necessary task than now. Even the conservative *New York Times* warned secessionists that among the likely consequences of their action *"the most unquestionable is War. . . .* [There] is no possibility of escaping it."[26]

President-elect Lincoln shared the determination of the "stiff-backed" Republicans to defend the Union, whatever the cost. Soon after the election he realized that stern measures might eventually be necessary. Recognizing southern independence would be a violation of his oath of office. The President derives his authority from the people, he said, and they had not empowered him to arrange the terms for a dissolution of the Union.[27]

In the past Lincoln had spoken eloquently in support of self-determination for Hungary and, more generally, of all people's "sacred" right of revolution. In 1848, while serving in Congress, he had said, "Any people anywhere, have the *right* to rise up, and shake off the existing government, and form a new one that suits them better." Moreover, that right is not limited to cases in which the whole people of a nation choose to exercise it. "Any portion of such people," he avowed, *"may* revolutionize, and make their *own* of so much of the territory as they inhabit."[28] In 1861, however, when contemplating domestic rather than foreign revolution, Lincoln qualified his position. "The right of revolution," he now claimed, "is never a legal right. . . . At most, it is but a moral right, when exercised for a morally justifiable cause. When exercised without such a cause revolution is no right, but simply a wicked exercise of physical power."[29] Lincoln thus viewed the southern rebellion as established governments have always viewed rebellion, whatever its cause—that is, as lacking the moral base required to give it validity. The true issue, he said, was not self-determination but whether "a democracy—a government of the people, by the same people—can, or cannot, maintain its territorial integrity against its own domestic foes."[30]

Lincoln was equally adamant in rejecting any compromise that involved a retreat from the platform on which he had been elected. "Hold firm, as with a chain of steel," he admonished his Republican friends in Congress. "The tug has to come, and better now, than any time hereafter." Concessions won by the threat of secession, he warned, would destroy the government.[31]

Looking back late in the war, Lincoln said, "I claim not to have controlled events, but confess plainly that events have controlled me."[32] That was only half true. He could not, of course, have controlled the South's radical response to his election, but his was the decisive voice in the development of the Union government's determined counterresponse. Given his administrative inexperience, Lincoln's clarity of vision, sureness of purpose, and skill in execution during the crisis were quite remarkable. While others in the North talked of peaceful secession, or of "masterly inactivity," or of compromise, sometimes wavering uncertainly between them, Lincoln carefully but resolutely prepared to do what he believed his oath of office required of him—that is, "to run the machine as it is."[33]

As soon as he heard of South Carolina's secession, he asked Lt. General Winfield Scott to be as well prepared as possible "to either *hold,* or *retake* the [southern] forts, as the case may require."[34] An Illinois friend recorded Lincoln's opinion that "far less evil and bloodshed would result from an effort to maintain the Union and the Constitution, than from disruption and the formation of two confederacies."[35] In February, while en route to Washington for his inauguration, the President-elect found an opportunity during one of his stops to assure the public that he did not propose to use the army to invade and coerce the southern states. But, he asked, would it be invasion or coercion merely to hold federal property and collect federal revenues? The Unionism of those who thought so he regarded as "of a very thin and airy character." In his inaugural address, he denied that he would be responsible for any violence that might ensue: "In *your* hands, my dissatisfied fellow countrymen, and not in *mine,* is the momentous issue of civil war. The government will not assail *you.* You can have no conflict without being yourselves the aggressors."[36] Defining coercion and aggression as he did, Lincoln could renounce them both and yet make clear his firm determination to enforce the laws and defend the government from aggressive acts against it. As one admiring Republican editor observed, Lincoln had hedged the secessionists in "so that they cannot take a single step without making treasonable war upon the government, which will only defend itself."[37]

After the inauguration, Lincoln wasted no time in preparing to implement the policy he was determined to pursue.[38] When he learned that the federal garrison at Fort Sumter in Charleston harbor was running short of supplies, he hesitated only because Major Robert Anderson, its commander, and General Scott advised him that supplying the garrison would require a larger military and naval force than could be mobilized in time to succeed. Early in April, in spite of their advice, as well as the urgent recommendation of Secretary of State Seward that Sumter be abandoned, Lincoln dispatched a relief expedition. Although it failed and the fort was lost, he had every reason to believe that even by the loss his policy had been well served. "You and I both anticipated," he wrote the commander of the expedition, "that the cause of the country would be advanced by making the attempt to provision Fort Sumpter [*sic*], even if it should fail, and it is no small consolation now to feel that our anticipation is justified by the result."[39] The Confederates had fired the first shot, and Lincoln now had a nearly united North behind him.

The resulting war engaged two similarly limited political democracies—limited, because all the Confederate states and all but a few of the loyal states gave the franchise only to white males. Both belligerents at the start set for themselves decidedly conservative goals. Jefferson Davis, in a message to the Confederate Congress, described the southern struggle for national self-determination as an effort to maintain an "indispensable" African slave-labor system and to protect "property worth thousands of millions of dollars."[40] Most Northerners in turn claimed to fight only to restore the old Union. In July, 1861, both houses of Congress resolved overwhelmingly that when the rebellion was suppressed the states would retain all their "dignity, equality, and rights . . . unimpaired," thus assuring Southerners that slave property would not be disturbed. "The Constitution as it is, and the Union as it was," soon became the slogan of conservative northern Democrats.

None was more determined than the nationalist Lincoln that the war should be waged solely to preserve the Union.

Although he had long hated slavery and had asserted that the Union could not endure permanently "half slave and half free," he had never advocated the overthrow of slavery by force. In the present crisis he believed that the cause of the Union must have priority over the cause of the slave. In his message to a special session of Congress, in July, 1861, Lincoln recalled and confirmed the pledge he had made in his inaugural address: "I have no purpose, directly or indirectly, to interfere with the institution of slavery in the States where it exists."[41] For the first two years of the war vindicating the principle of a perpetual Union remained his single goal.

In contrast, militant abolitionists always professed a dual purpose. "We mean both *Emancipation* and *Union,*" wrote one, "the one for the sake of the other and both for the sake of the country." Neal Dow, the Maine abolitionist, predicted that northern soldiers would not return from their mission "until the question of slavery should be settled forever." Once the war began even the Garrisonians abandoned their pacifism and resolved to turn the war into an antislavery crusade. "I speak under the stars and stripes," the Boston abolitionist Wendell Phillips now affirmed, for "to-day the slave asks God for a sight of this banner, and counts it the pledge of his redemption."[42]

Month by month, with no end to the war in sight, the ranks of those who demanded the destruction of slavery increased, and the Republican majority in Congress began to act. In December, 1861, the House of Representatives refused to renew its pledge of the previous July not to interfere with slavery. In 1862 Congress abolished slavery in the District of Columbia and prohibited it in the territories. Two confiscation acts sought, with only limited success, to emancipate southern slaves used for military purposes and those owned by disloyal masters.[43]

Lincoln, however, refused to budge. Instead he pressed his

own conservative program of gradual, voluntary emancipation and the colonization of free blacks abroad. He urged Congress to give financial assistance to any state that would adopt a program of gradual emancipation. Although he was disappointed in the result, his aim was to win the cooperation of the border slave states and thus deprive the Confederacy of the hope that they would eventually secede.[44] Clearly, Lincoln was not shifting his ground but merely pursuing his same fixed goal by other means.

By the summer of 1862 some radical Republicans were outspokenly impatient with his stubborn conservatism, and abolitionists attacked him bitterly. Lincoln's actions, the black abolitionist Frederick Douglass complained, had been calculated to protect the property of slaveholders, and his policy was "to reconstruct the union on the old and corrupting basis of compromise." "He is nothing better than a wet rag," fumed Garrison. In the opinion of Wendell Phillips, he was conducting the war "with the purpose of saving slavery." The present conflict was "aimless . . . wasteful and murderous. Better that the South should go to-day, than that we should prolong such a war." To Lincoln the most painful attack came from Horace Greeley. In an open letter to him published in the *New York Tribune*, Greeley accused the President of "mistaken deference" to rebel slaveholders and to the "fossil politicians" of the border slave states. Every "disinterested, determined, intelligent champion of the Union," he asserted, believed that "all attempts to put down the Rebellion and at the same time uphold its inciting cause are preposterous and futile."[45]

The transparently angry President was uncharacteristically sharp in his reply. He had not meant to leave any doubt about his policy: "My paramount object in this struggle *is* to save the Union, and is *not* either to save or to destroy slavery." To advance that cause he was ready to free all of the slaves, or some of them, or none. "What I do about slavery, and the colored race, I do because I believe it helps to save the Union; and what I forbear, I forbear because I do *not* believe it would help to save the Union." Lincoln concluded by repeating his genuine and "oft-expressed *personal* wish that all men everywhere could be free."[46]

The fact that Lincoln had decided a month before his reply to Greeley to issue an Emancipation Proclamation did not mean that he was secretly pursuing one policy and publicly another. In his Preliminary Proclamation of September 22, 1862, he announced that "hereafter, as heretofore," the war would be prosecuted to preserve the Union, and that he would continue to press for voluntary, gradual, compensated emancipation and colonization. Lincoln justified the final Emancipation Proclamation of January 1, 1863, as "a fit and necessary war measure" for suppressing the rebellion. In applying it to the states or portions of states still in rebellion he described it as "an act of justice, warranted by the Constitution, upon military necessity." The remark of one historian that the Proclamation "had all the moral grandeur of a bill of lading" was descriptively accurate; but it overlooked the mitigating fact that Lincoln was offering a constitutional justification for his action and seeking maximum support from conservative Unionists who wanted no part in an abolitionist crusade.[47]

Even so, the Proclamation, which also authorized the recruitment of blacks for the Union army, did less than justice to an act potentially so momentous in its social consequences. Apart from political expediency, the reason, in all probability, was that when Lincoln issued it he himself did not fully recognize that the conflict thereby would be transformed into a great social revolution. In his view, it was still a war for the Union, nothing more. "For this alone have I felt authorized to

Dictator Lincoln. TOP: The ca. 1864 etching of Adalbert Volck of Baltimore showed Lincoln writing the Emancipation Proclamation with his foot on the American Constitution, the Devil holding his inkstand, and Liberty's face covered with a hood in the upper left corner. *Courtesy of Gettysburg College Special Collections.* BOTTOM: In John Tenniel's 1863 *London Punch* cartoon Lincoln meets his equal in the Czar of Russia, Alexander II. Both were putting down rebellions at the time, one in the American South, the other in Poland. *Courtesy of the* Punch *Library, London*.

struggle," he assured a critic, "and I seek neither more nor less now."[48]

What, then, caused Abraham Lincoln, the nationalist, the narrowly focused, almost obsessive defender of the Union during the war's first two years, to broaden his vision and become at last the Great Emancipator? It was hardly a role that he had anticipated. This remarkable transformation began sometime in the summer of 1863. By then the war had gone on too long, its aspect had become too grim, and the escalating casualties were too staggering for a man of Lincoln's sensitivity to discover in that terrible ordeal no greater purpose than the denial of the southern claim to self-determination. The great battles of the spring and summer of 1862—Shiloh, the Seven Days, Second Bull Run, and Antietam—had brought home to him the magnitude of the task he had undertaken. The combined Union and Confederate casualties in those four battles and in the four that followed—Fredericksburg in December, Chancellorsville in May, 1863, Gettysburg in July, and Chickamauga in September—numbered 234,000. Proportionally, in our present population, the casualties of those eight battles, fought in a period of eighteen months, would have amounted to nearly two million.

Sharing responsibility for the events that had brought these lamentable results was more than Lincoln had bargained for when he won the presidency. As Richard Hofstadter observed in a perceptive biographical essay: "Lincoln was moved by the wounded and dying men, moved as no one in a place of power can afford to be. . . . For him it was impossible to drift into the habitual callousness of the sort of officialdom that sees men only as pawns to be shifted here and there and 'expended' at the will of others."[49] Bearing this heavy burden, being a man of deeply religious temperament, it was natural for him, amid the death and destruction, to search for a divine purpose, one that perhaps he had failed to comprehend.

An early indication of Lincoln's broadening conception of the war's meaning was his response to a serenade a few days after the Union victory at Gettysburg. He was not then prepared, he said, to deliver an address worthy of the occasion, but he spoke briefly and feelingly of the need to defend the principle that "all men are created equal" against those who would subvert it.[50] He returned to that theme in his memorable Gettysburg Address, in which, near the end, another sign of his nascent vision appeared. When Lincoln expressed the hope that those who died at Gettysburg "shall not have died in vain—that this nation, under God, shall have a new birth of freedom," he was, for the first time, anticipating the imminent end of slavery as well as the preservation of the Union.[51] It is reasonable, I think, to give an abolitionist meaning to his phrase "a *new* birth of freedom."

By the time of his Gettysburg Address, Lincoln had abandoned his public posture of indifference to the fate of southern slaves manifested in his letter to Horace Greeley the previous year. On several occasions, including his third and fourth annual messages to Congress, he vowed that he would not "retract the emancipation proclamation; nor, as executive, ever return to slavery any person who is free by the terms of that proclamation, or by any of the acts of Congress." Could such treachery, he asked, "escape the curses of Heaven, or of any good man?"[52] By the summer of 1864, after the 86,000 Union and Confederate casualties of General Grant's Wilderness campaign, the change in Lincoln's vision was complete, for he would no longer make peace merely on the basis of a restored Union. Fully aware of the contribution of black troops to the Union cause, he now insisted that an acceptable peace must include both "the integrity of the whole Union, and the abandonment of slavery."[53]

In June, 1864, when the Thirteenth Amendment, providing final and complete emancipation, first came to a vote in the House of Representatives, it failed to win the required two-thirds majority. Lincoln's wholehearted support was crucial in getting that vote reversed the following January. Responding to a serenade after the passage of the amendment, he congratulated the country "upon this great moral victory."[54] Lincoln had indeed become the Great Emancipator, and William Lloyd Garrison concluded that he was more than a "wet rag" after all. In a letter, dated February 13, 1865, Garrison commended him warmly for the part he had played in the final abolition of slavery: "As an instrument in [God's] hands," he wrote, "you have done a mighty work for the freedom of millions . . . in our land. . . . I have the utmost faith in the benevolence of your heart, the purity of your motives, and the integrity of your spirit."[55]

An instrument in God's hands. That seemed to be the role to which Lincoln had resigned himself when he delivered his beautiful and deeply moving second inaugural address. In this, his final effort to grasp the meaning of the war, he came full circle, for the cause of the Union now seemed ancillary to the approaching liberation of four million slaves. Perhaps it was God's will, he suggested, that the war must continue "until all the wealth piled by the bond-man's two hundred and fifty years of unrequited toil shall be sunk, and until every drop of blood drawn with the lash, shall be paid by another drawn with the sword."[56] Lincoln spoke without malice. In his view the cause of the slave was beyond malice, the guilt of slavery was shared by all, and retribution was best left to divine judgment.

The profoundly religious sentiments expressed in the second inaugural address were those of a man who not only had led the nation through a devastating crisis but, because of it, had experienced an unsettling personal crisis as well. The address was his testament, his witness, that, by embracing the cause of the slave, he had found the war's ultimate justification and thereby a way to come to terms with his God and with himself. Nothing less personal could explain the depth of feeling that his words revealed.

Reading Lincoln's private letters and public papers from 1848 to 1865 leaves one with the impression that on the issue of self-determination his legacy to posterity is both ambiguous and complex. In spite of his earlier defense of that principle, his nationalism and belief in the perpetuity of the Union had led him to crush the one attempt in our history to apply it at home. Secession, he affirmed, was neither a constitutional procedure nor an appropriate extralegal remedy for alleged grievances in a democracy such as ours.

That was part of his legacy. But he left unanswered the question of when, by whom, and for what cause the right of self-determination could be justifiably invoked. That question still remains unanswered. Being an internal matter, it has never been treated in international law. Woodrow Wilson's Fourteen Points only dealt with specific cases, and his more general statements were vague to say the least. The United Nations Charter commits its members to respect the principle of self-determination, but it, too, fails to establish standards by which to judge the claims of the many ethnic groups who assert their right to independence.[57]

Historically, the success of movements for self-determination has had little to do with the justice or morality of individual cases. Success has depended on the good will of the national state involved, as may eventually be the case of Quebec; or on its inherent weakness, which has been the case of the remaining republics in the Soviet Union; or on the decisions of victorious great powers, as were the cases of Poland, Czechoslovakia, and Yugoslavia after the first World War; or on the military strength of the rebels themselves, as was the case of the United States, in alliance with France, in 1783.

In 1918, Robert Lansing, Woodrow Wilson's Secretary of State, expressed skepticism about the very principle of self-determination, believing that it was unworkable and full of mischief. Recently, Arthur Schlesinger, Jr., warned that the current widespread assertion of the principle is a potential threat to the unity of most national states, and that even in the United States "the outburst of multicultural zealotry threatens . . . a new tribalism." Moreover, Lansing wondered, if self-determination were a valid principle, how can we justify Lincoln's refusal to grant independence to the southern Confederacy? [58]

However, Lincoln ultimately escaped that dilemma by attributing Confederate defeat to divine intervention on the side of a just and moral cause. In the end, when the cause of the Union no longer seemed to be sufficient, he invoked the cause of the slave, rather than the authority of the Constitution, to justify the sacrifice of so many lives. That is what makes his legacy ambiguous.

Nevertheless, if self-determination should ever again become an internal issue for the United States, it would be quite logical for us to turn to Lincoln's messages and papers for guidance. Among them we would find not only the clear imperative of his militant first inaugural address but the chastening words of his second inaugural address as well. Let us be sure, if such a time should come, that the foundation of the Union cause will be at least as just and moral as his!

6

War and the Constitution: Abraham Lincoln and Franklin D. Roosevelt

ARTHUR M. SCHLESINGER, JR.

*O*DDLY, THE GETTYSBURG ADDRESS made no great impression on November 19, 1863. Lincoln's bright and devoted young secretary John Hay casually noted in his diary: "The President, in a fine, free way, with more grace than is his wont, said his half dozen words of consecration, and the music wailed, and we went home through crowded and cheering streets. And all the particulars are in the daily papers."[1] It took time for the Gettysburg Address to become a classic statement of the American creed. Today one sometimes feels that Lincoln's crystalline words have grown so familiar that they are part of the mechanical ritual of our lives—words we hear and repeat but no longer attend to.

The more venerable among us may remember *Ruggles of Red Gap,* a 1935 film in which Charles Laughton, playing an English butler won in a poker game by an American rancher, electrifies his new employers—and movie audiences of the 1930s sitting in darkened theaters—by remembering the Gettysburg Address, when his new boss had forgotten it, and delivering it as if each stunning phrase had come fresh from his mind. Laughton made us listen anew and made us think anew. For the testing of which President Lincoln so wonderfully spoke— whether a nation conceived in liberty and dedicated to the proposition that all men are created equal can long endure— was not only the aspiration of 1776 and the challenge of 1863 but must remain the unending commitment of all Americans till the end of our days.

The republic has gone through two awful times of testing since the achievement of independence—two times when the life of the nation was critically at stake, two times when the nation was led by Presidents absolutely determined that government of the people, by the people, for the people, should not perish from the earth. The two Presidents had to confront the question whether the Constitution of 1787 was equal to the cruel emergencies of 1861 and 1941. History in these periods subjected our republican institutions to their severest trials— and, with the survival of the Constitution, saw their greatest triumphs.

The two Presidents were very different men in very different situations. Abraham Lincoln striding from the backwoods of the middle border was the common man incarnate. Franklin D. Roosevelt was a Hudson River patrician. Lincoln was self-educated. Roosevelt received the best education his country could provide. Lincoln was chosen to head a republic of thirty-four states with a population of thirty-two million; Roosevelt a republic of forty-eight states with a population four times as great. Lincoln faced a civil war within the United States; Roosevelt, a foreign war threatening to engulf the planet. Lincoln operated a presidency of still largely undefined powers; Roosevelt, a presidency considerably more secure in its assertions of national leadership. Lincoln was enveloped by a tragic sense of life. Roosevelt breezed through life in confident and incurable optimism.

Yet they had many similarities. Both were men of mysterious and impenetrable reserve, concealing resolute purpose behind screens of fable, parable, and jocosity. Both had deep and irreversible moral convictions about freedom and human rights. Both were skilled, crafty, and, when necessary, ruthless politicians. Both were lawyers who, while duly respecting their profession, regarded law as secondary to political leadership. Both had faith that the Constitution, spaciously interpreted, could surmount crisis. Lincoln, following his hero Henry Clay, "my beau ideal of a statesman," in a broad reading of the national charter, said in his First Inaugural: "I take the official oath today, with no mental reservations, and with no purpose to construe the Constitution or laws, by any hypercritical rules."[2] Roosevelt, following his heroes Theodore Roosevelt and Woodrow Wilson in their robust conceptions of presidential leadership, said in his First Inaugural, "Our Constitution is so simple and practical that it is possible always to meet extraordinary needs by changes in emphasis and arrangement without loss of essential form."[3]

And both Presidents confronted national emergencies that demanded bold and peremptory action. Both assumed powers that led other Americans to charge that the Constitution had been transgressed and betrayed. Both provoked cries of dictatorship. Both, in responding to what they saw as the necessities of the day, risked the creation of dangerous precedents for the future. An examination of the manner in which Lincoln and Roosevelt met their emergencies may illuminate our understanding of the potentialities, limits, and perils of presidential leadership. I propose to discuss in particular the way these two Presidents handled first the war-making power and then threats to internal security once war had begun.

The men who drafted the Constitution in Philadelphia in 1787 had questions of national defense much on their minds. The remedy for the infant nation's international vulnerabilities was, they believed, a strong central government empowered to create a standing army and navy, to regulate commerce, to enforce treaties and, when necessary, to go to war. But, given the separation of powers, how should foreign policy authority be distributed within the new national government?

Here the Framers were unambiguous in their decisions. The vital powers were to be reserved for Congress. Article I of the new Constitution gave Congress not only the exclusive appropriations power—itself a potent instrument of control—but the exclusive power to declare war, to raise and support armies, to provide and maintain a navy, to make rules for the government and regulation of the armed services, and to grant letters of marque and reprisal—the last provision representing the eighteenth-century equivalent of retaliatory strikes and enabling Congress to authorize limited as well as formal war. Even Alexander Hamilton, the convention's foremost proponent of executive energy, endorsed this allocation of powers, expressly rejecting the notion that foreign policy was the private property of the President. "The history of human conduct," Hamilton wrote in the 75th Federalist, "does not warrant that exalted opinion of human virtue which would make it wise in a nation to commit interests of so delicate and momentous a kind, as those which concern its intercourse with the rest of the world, to the sole disposal of a magistrate created and circumstanced as would be a President of the United States."

No one can doubt the determination of the Framers, in the

words of James Wilson, to establish a procedure that "will not hurry us into war; it is calculated to guard against it. It will not be in the power of a single man, or a single body of men, to involve us in such distress."[4] Sixty years later, during the Mexican War, Congressman Abraham Lincoln of Illinois accurately expressed original intent when he wrote that the convention "resolved so to frame the Constitution that *no one man* should hold the power of bringing this oppression upon us."[5]

While reserving decisive foreign policy powers for Congress, the Framers did assign the executive a role in the conduct of national security affairs. Instead of giving Congress the exclusive power to "make" war, as the draft under consideration stipulated, James Madison moved to replace "make" by "declare" in order to leave "to the Executive the power to repel sudden attacks."[6]

The President, moreover, was constitutionally designated commander in chief of the armed services. The Framers saw this, however, as a ministerial function, not as a grant of independent executive authority. The designation, as Hamilton wrote in the 69th *Federalist*, "would amount to nothing more than the supreme command and direction of the military and naval forces." It meant only that the President should have the direction of war once authorized or begun. Hamilton contrasted this limited assignment with the power of the British king—a power that "extended to the declaring of war and to the raising and regulating of fleets and armies,—all which by the Constitution under consideration, would appertain to the legislature."

The Constitution, in short, envisaged a partnership between Congress and the President in the conduct of foreign affairs with Congress as the senior partner. Yet one may suppose that another consideration lingered in the Framers' innermost thoughts—a fallback position that, in acknowledging the

hard realities of a dangerous world, justified a measure of unilateral executive initiative. This was the question of emergencies.

The Framers had been reared on John Locke. They were
well acquainted with chapter fourteen, "Of Prerogative," in
Locke's *Second Treatise on Civil Government*. While in normal
times, Locke said, responsible rulers must observe the rule of
law, in dire emergencies they could initiate extralegal or even
illegal action. Sometimes "a strict and rigid observation of the
laws may do harm." The executive, Locke contended, must have
the reserve power "to act according to discretion for the public
good, without the prescription of law and sometimes even
against it." The test of whether prerogative was rightfully invoked, Locke said, was whether the emergency was a true one
and whether the exercise of prerogative tended "to the good
or hurt of the people"—judgments to be made in the end not
by the ruler but by the people.

Prerogative was the application of the law of national self-
preservation. The doctrine was not conceded in the Constitution, except for a solitary provision permitting the suspension
of the writ of habeas corpus "when in Cases of Rebellion or
Invasion the public safety may require it" (Article I, section 9).
The limitation of emergency prerogative to rebellion and invasion and to the single matter of habeas corpus implied less a
standard that could be extended to other issues than a rejection
of any broader suspension of constitutional guarantees even in
emergencies.

Yet the notion that crisis might require the executive to act
outside the Constitution in order to save the Constitution still
lurked in the back of the minds of the men who won American
independence. Hamilton wrote in the 28th Federalist of "that
original right of self-defence which is paramount to all positive

forms of government"; and Madison in the 41st thought it "vain to oppose constitutional barriers to the impulse of self-preservation." Even Jefferson, the apostle of strict construction, affirmed the need for emergency prerogative. "On great occasions," he wrote in 1807, "every good officer must be ready to risk himself in going beyond the strict line of the law, when the public preservation requires it." There were, he said, "extreme cases where the laws become inadequate to their own preservation, and where the universal recourse is a dictator, or martial law."[7]

Nor was this a passing thought. Jefferson restated the point more fully after he left the White House. "A strict observance of the written laws," he wrote carefully in 1810, "is doubtless *one* of the high duties of a good citizen, but it is not *the highest*. The laws of necessity, of self-preservation, of saving our country when in danger, are of a higher obligation. . . . To lose our country by a scrupulous adherence to written law, would be to lose the law itself, with life, liberty, property and all those who are enjoying them with us; thus absurdly sacrificing the end to the means." He understood the risks in this argument and therefore placed emergency power under the judgment of history: "The line of discrimination between cases may be difficult; but the good officer is bound to draw it at his own peril, and throw himself on the justice of his country and the rectitude of his motives."[8]

Jefferson's defense of Lockean prerogative was inspired by his passion to protect the republic against Aaron Burr. This was a doubtful case. No one can be sure what the Burr conspiracy was up to. The House of Representatives, in voting down a proposal for the suspension of habeas corpus, rejected any idea that the life of the nation was at risk. Neither Jefferson's contemporaries nor future historians have been convinced that Jefferson faced an emergency so imperative as to justify a laying aside of the law. Burr's acquittal by the courts helped limit subsequent resort to emergency prerogative.

It was not enough for a President personally to think the country was in danger. To confirm a judgment of dire emergency, a President had to have the broad agreement of Congress and public opinion. Emergencies considerably more authentic than the Burr conspiracy took place in the next thirty years. But Presidents as forceful as Jackson and Polk refrained from invoking emergency prerogative—even in face of the nullification crisis and the war with Mexico.

Jefferson himself had restricted prerogative to "great occasions." But in fact he was prepared to ignore Congress and to take unilateral action on lesser occasions as well. Thus he sent a naval squadron to the Mediterranean under secret orders to fight the Barbary pirates, applied for congressional sanction six months later, and then misled Congress as to the nature of the orders. He unilaterally authorized the seizure of armed vessels in waters extending to the Gulf Stream, engaged in rearmament without congressional appropriations, and not infrequently withheld information from Congress.

Others of our early Presidents imitated Jefferson's unilateral initiatives. As Judge A. D. Sofaer has shown in his magistral work, *War, Foreign Affairs and Constitutional Power: The Origins,* unauthorized presidential adventurism thrived in the early republic. But these Presidents did not assert it as their constitutional right to ignore Congress and strike out on their own. "At no time," Sofaer writes of the classical period, "did the executive claim 'inherent' power to initiate military action."[9] Sofaer's surmise is that our early Presidents deliberately selected venturesome agents, deliberately kept their missions secret, deliberately gave them vague instructions, deliberately declined to approve or disapprove their constitutionally questionable plans, and deliberately denied Congress the information to determine whether aggressive acts were authorized—all precisely because the Presidents wanted to do things they knew lay beyond their constitutional right to command.

The partnership between Congress and the executive in the conduct of foreign affairs was thus unstable from the start. President Polk, in getting into a war with Mexico that, according to Lincoln, had been unnecessarily and unconstitutionally begun, showed both the potentialities of presidential power and the limitations of legislative control. Despite his own strong opposition to the Mexican War, Lincoln had the advantage of Polk's vigorous example when he returned to Washington a dozen years later, now President himself, facing not foreign war but domestic insurrection.

Domestic insurrection raised a different set of constitutional issues, and this simplified Lincoln's problem. He did not—or at least so he believed—need congressional recognition of a state of war, as he would have done against a foreign state (four Supreme Court justices soon opined otherwise, however, in the Prize cases). He had only, he believed, to carry out his presidential duty of enforcing domestic law against rebellious individuals.

Still even this duty implied in the circumstances a warlike course that might well call for congressional approval. And that warlike course called for auxiliary measures that certainly required congressional action. Lincoln chose nevertheless to begin by assuming power to act independently of Congress. Fort Sumter was attacked on April 12, 1861. On April 15, Lincoln summoned Congress to meet in special session—but not till July 4. He thereby gained ten weeks to bypass Congress, rule by decree, and set the nation irrevocably on the path to war.

On April 15, he called out state militia to the number of seventy-five thousand. Here he was acting on the basis of a statute. From then on he acted on his own. On April 19, he imposed a blockade on rebel ports, thereby assuming authority to take actions hitherto considered as requiring a declaration of war. On May 3, he called for volunteers and enlarged the army and navy, thereby usurping the power confided to Congress to raise armies and maintain navies. On April 20, he ordered the Secretary of the Treasury to spend public money for defense without congressional appropriation, thereby violating Article I, section 9, of the Constitution. On April 27, he authorized the commanding general of the army to suspend the writ of habeas corpus—this despite the fact that the power of suspension, while not assigned explicitly to Congress, lay in that article of the Constitution devoted to the powers of Congress and was regarded by commentators before Lincoln as a congressional prerogative. Later he claimed the habeas corpus clause as a precedent for wider suspension of constitutional rights in time of rebellion or invasion—an undoubted stretching of original intent.

When Congress finally assembled on July 4, Lincoln justified his actions. The issue, he said, embraced more than the fate of these United States. The rebellion forced "the whole family of man" to ask questions going to the roots of self-government: " 'Is there in all republics, this inherent and fatal weakness?' 'Must a Government, of necessity, be too *strong* for the liberties of its own people, or too *weak* to maintain its own existence?" So viewing the issue, Lincoln continued, "no choice was left but to call out the war power of the Government; and so to resist force employed for its destruction, by force for its preservation." [10]

The phrase "war power" was novel in American constitutional discourse. John Quincy Adams, it is true, had contrasted in 1836 the "peace power" as something limited by the Constitution as against the "war power," limited only by the laws and usages of nations.[11] But Adams was speaking about the war power of the national government as a whole, exercised through and with Congress. He was not speaking, as Lincoln was, about the war power as a peculiar function of the executive.

The "war power" flowed into the presidency, as Lincoln saw it, through the presidential oath to "preserve, protect, and defend the Constitution," through the constitutional commitment to take care that the laws be faithfully executed, and through the constitutional designation of the President as commander in chief. "I think," he later said, "the Constitution invests its commander-in-chief, with the law of war, in time of war"[12]—a statement that, if not altogether clear, was certainly pregnant. It must be noted, however, that Lincoln limited that investment of power to wartime, thereby excluding twentieth-century tendencies to argue that the clause bestows powers on the presidency in times of peace.

Still, Lincoln's reading of the clause greatly enlarged presidential power in war. His most far-reaching action, the Emancipation Proclamation, began by invoking "the power in me vested as Commander-in-Chief of the Army and Navy" and ended by justifying emancipation as "warranted by the Constitution, upon military necessity." He later characterized the Proclamation as without "constitutional or legal justification, except as a military measure." He added: "I conceive that I may in an emergency do things on military grounds which cannot be done constitutionally by Congress." And: "As commander-in-chief of the army and navy, in time of war, I suppose I have a right to take any measure which may best subdue the enemy."[13]

Lincoln did not himself define the limits of the executive war power. An exculpatory opinion, extracted from his somewhat reluctant Attorney General, Edward Bates, contended that the national emergency justified Lincoln in suspending habeas corpus and disregarding subsequent judicial objection, even from so august a source as Chief Justice Roger Taney in *Ex parte Merryman*. The President, Bates added, was the judge of the gravity of the emergency and was accountable only through procedures of impeachment.

But Lincoln, though he had begun by acting without congressional authorization, had no intention of ruling Congress out of the game. His actions, he told Congress when it finally assembled, "whether strictly legal or not, were ventured upon under what appeared to be a popular demand and a public necessity; trusting then as now that Congress would readily ratify them. It is believed that nothing has been done beyond the constitutional competency of Congress."

It was necessary to suspend habeas corpus, Lincoln added, in order to assure the enforcement of the rest of the law and thereby the protection of the state. "Are all the laws *but one* to go unexecuted, and the Government itself go to pieces, lest that one be violated? . . . In such a case, would not the official oath be violated if the government should be overthrown?" Would the very principles of freedom prevent free government from defending itself? As Lincoln explained his case toward the end of the war, his oath to preserve the Constitution imposed the "duty of preserving, by every indispensable means, that government—that nation—of which the constitution was the organic law. Was it possible to lose the nation, and yet preserve the Constitution?"[14]

Lincoln took that duty with utmost seriousness and assessed the internal threat behind the lines in the North with stern urgency. Rebel sympathizers, he said, "pervaded all departments of the government and nearly all communities of the people. . . . Under cover of 'Liberty of speech,' 'Liberty of the press,' and *'habeas corpus,'* they hoped to keep on foot amongst us a most efficient corps of spies, informers, supplyers, and aiders and abettors of their cause in a thousand ways." Conspiracy-mongers like the detectives La Fayette Baker and Allan Pinkerton inflamed the official imagination. Northern opponents of the war were denounced as Copperheads. Invoking his "war power," Lincoln set in motion a series of drastic though often inept *ad hoc* actions: martial law and military courts, some far from the fighting fronts; detectives (the twentieth century learned to call them secret police) and paid informers; military arrest and detention of untold thousands of persons; suppression of newspapers; seizure of property; denial of the mails to "treasonable correspondence"—all in the belief that "certain proceedings are constitutional when, in cases of rebellion or Invasion, the public Safety requires them, which would not be constitutional when, in absence of rebellion or invasion, the public Safety does not require them."[15]

Such actions, though tempered by Lincoln's restraint and humanity, provoked denunciations of despotism and cries of dictatorship. In 1862 the eminent lawyer Benjamin R. Curtis, who five years earlier had been a dissenting Supreme Court justice in the Dred Scott case and six years later would be Andrew Johnson's counsel in the impeachment proceedings, published a cogent pamphlet condemning Lincoln's proclamations and orders as "assertions of transcendent executive power" having the effect of placing "every citizen of the United States under the direct military command and control of the President."[16]

The exuberant Secretary of State, W. H. Seward rejoiced in the situation. "We elect a king every four years," he told the London *Times* correspondent, "and give him absolute power within certain limits, which after all he can interpret for himself."[17] If anything, Secretary of War Edwin Stanton rejoiced more. Even so measured a commentator as Lord Bryce could write in a few years that Lincoln was "almost a dictator . . . who wielded more authority than any single Englishman has done since Oliver Cromwell."[18] The Civil War, Henry Adams wrote five years after Appomattox, "for the time obliterated the Constitution."[19]

Of course Lincoln was far from a dictator. The mechanisms of accountability—Congress, the courts, free elections, substantial freedoms of speech, press and assembly—all remained in place. No dictator would have tolerated such fierce opposition in Congress and such bitter criticism in the newspapers. Nor would a dictator have submitted to a presidential election in the midst of war—and made preparations, in case he lost, to cooperate with his successor. Nor would a dictator have tolerated a Copperhead as Chief Justice of the Supreme Court. Lincoln did not even seek a Sedition Act of the sort Congress had given the executive in 1798, or an Espionage Act, as in 1917.

Still, in issuing decrees without legislative authorization, Lincoln assumed quasi-dictatorial powers. And no doubt he exaggerated the internal threat to national security. But civil wars are desperate affairs. The North did in fact have many persons opposed to the war. Some Copperheads were in fact Confederate agents. A responsible President could not afford to take chances. One might wish that Lincoln had acted at the time with the wisdom available to historians after the peril had passed. But Lincoln had to reckon with the gravest threat to the life of the republic, and he could not foretell the outcome. "It is very difficult to remember," wrote Maitland, "that events now in the past were once in the future."[20] We know how it all came out. Lincoln did not.

As usual, Lincoln found the homely analogy to defend his course. Human beings, he observed, wished to protect life and limb. "Yet often a limb must be amputated to save a life; but a life is never wisely given to save a limb. I felt that measures, otherwise unconstitutional, might become lawful by becoming indispensable to the preservation of the constitution, through

The Federal Phoenix. Like the fabled bird of ancient Egyptian mythology, the re-elected Lincoln rises from ashes of American liberties and prosperity: credit, free press, *habeas corpus,* and more. The cartoon by John Tenniel appeared in the London *Punch,* in 1864. *Courtesy of the* Punch *Library, London*.

the preservation of the nation."[21] One recalls Jefferson's point about absurdly sacrificing the end to the means; one hears the Lockean echo, even though the Locke to whom Lincoln most often referred was the lesser Locke who wrote under the name of Petroleum V. Nasby.

Lincoln secured congressional ratification of most of his unilateral actions. Such ratification might be taken as legislative obeisance to an imperial President—or as legislative affirmation that, despite the emergency, Congress retained its constitutional powers. With the war still on, a divided Supreme Court in 1863 in the Prize cases rejected the contention that those actions Lincoln took unilaterally before Congress ratified them represented merely his "personal war against the rebellion." The majority ruled that the attack on Fort Sumter created a *de facto* state of civil war and that the President was "bound to meet it in the shape it presented itself, without waiting for Congress to baptize it with a name."[22] In the stress of war the judiciary too accepted what the executive had ventured upon under a popular demand and a public necessity.

The emergency Franklin Roosevelt faced eighty years later assumed a different form but presented almost as mortal a threat to the life of the nation. By the summer of 1940, Great Britain stood alone against Hitler. With nearly half the British destroyer fleet sunk or damaged, with a Nazi invasion of Britain darkly in prospect, Winston Churchill, the new British prime minister, asked Roosevelt for the loan of American destroyers "as a matter of life and death."[23] Weighing this anguished request, Roosevelt, for all his desire to aid Britain, was acutely mindful of the constitutional role of Congress.

When the French prime minister had asked earlier that spring for American assistance against the Nazi blitzkrieg, Roosevelt had replied that, while the United States would continue supplies so long as the French continued resistance, "I know that you will understand that these statements carry with them no implications of military commitments. Only the Congress can make such commitments."[24] To Churchill's plea for the loan of destroyers, Roosevelt initially responded that "a step of this kind could not be taken except with the specific authorization of the Congress, and I am not certain that it would be wise for that suggestion to be made to the Congress at this moment."[25] Not only would such a step enrage the isolationist opposition in Congress but it was also an explosive issue to throw into the 1940 presidential campaign. As late as August of this dangerous year Roosevelt continued to believe that a transfer of destroyers would require congressional action.

In the meantime, able New Deal lawyers, notably Benjamin V. Cohen and Dean Acheson, construed the applicable statutes to mean that unilateral executive action would be legal if the transfer of destroyers could be shown to strengthen rather than to weaken American defenses. At first the President heard the new argument with skepticism. But the British plight grew more desperate and Churchill's pleas more urgent. Roosevelt now moved with careful, if informal, concern for the disciplines of consent. He consulted his cabinet. He consulted congressional leaders. Through intermediaries he consulted the Republican candidates for President and Vice President, Wendell Willkie and Senator Charles McNary. McNary, a public-spirited man, was also the Republican leader of the Senate, and he soon passed word to the White House that, while it would be hard for him to vote for a statute authorizing the transfer of destroyers, he would make no objection if persuasive grounds could be found for going ahead without resort to Congress.

Roosevelt then extracted from his somewhat reluctant Attorney General, Robert H. Jackson, an opinion telling him that he could by executive agreement exchange destroyers for bases in British possessions in the Western Hemisphere. Jackson mentioned the commander in chief clause only to note, "Happily, there has been little occasion in our history for the interpretation of the powers of the President as Commander in Chief." Instead of relying upon the "constitutional power" of the presidency, Jackson found "ample statutory authority to support the acquisition of these bases." His opinion rested basically on a construction of laws passed by Congress, not on theories of inherent executive authority. Later Jackson observed that Roosevelt "did not presume to rely upon any claims of constitutional power as Commander-in-Chief" but made the transfer because, as he read the law, "Congress so authorized him."[26]

Critics thought the Attorney General's opinion strained, and Jackson himself years later made a semi-disclaimer. The great constitutional scholar E. S. Corwin called the opinion at the time "an endorsement of unrestrained autocracy in the field of our foreign relations," adding hyperbolically that "no such dangerous opinion was ever before penned by an Attorney General of the United States."[27]

Even great constitutional scholars can overreact, and in this case Corwin surely overreacted. The Jackson opinion was a response to a unique emergency; it received tacit congressional ratification when Congress appropriated money to build the bases; and to my knowledge it has never since been cited as justification for solo executive exploits in foreign affairs. The destroyer deal was compelled by a threat to the republic surpassed only by the emergency Lincoln faced after Sumter. It seems less a flagrant exercise in presidential usurpation than a defensible application of the Locke-Jefferson-Lincoln doctrine of emergency prerogative.

The destroyer deal was an unneutral act. Still, as international lawyers pointed out at the time, Hitler's own scorn for neutral rights weakened any claim he might make for the neutral rights of Nazi Germany. The deal did not (as some have said in recent years) violate domestic neutrality legislation. That legislation governed economic, not political, relations between the United States and belligerent states. It prohibited loans, credits, arms sale, and travel under specified conditions; it did not prohibit choosing sides.

The really decisive step away from neutrality, however, was not taken unilaterally by the President. It was taken with due solemnity by the President and Congress together in March 1941. Instead of relying on inherent presidential power, Roosevelt asked Congress to enact the Lend-Lease bill, a bill that, if it became law, would align the United States in the most unequivocal manner with Britain in its war against the Axis states. After two months of vigorous debate, Congress passed Lend-Lease by comfortable margins in both houses.

The Lend-Lease Act set the course for the months that followed. As Cordell Hull, the Secretary of State, told the American Society of International Law in April, the declared policy of the legislative and executive branches to give aid to Britain "means in practical application that such aid must reach its destination in the shortest of time and in maximum quantity. So ways must be found to do this."[28]

Once Congress had authorized the lending and leasing of goods to keep Britain in the war, did this authorization not imply an effort to make sure that the goods arrived? So Roosevelt assumed, trusting that a murky proclamation of "unlimited national emergency" in May and the impact of Nazi aggression on public opinion would justify his policy. In protecting the British lifeline, Roosevelt in the next months undertook a series of steps that by autumn had thrust the United States into an undeclared naval war in the North Atlantic. These steps—U.S. naval patrols that soon turned into convoys half-way across the ocean; the despatch of American troops to Greenland and soon to Iceland; cooperation with the British navy in tracing and sinking German U-boats; misrepresentation of the German attack on the destroyer *Greer;* the shoot-at-sight policy in patrol zones in September—were taken on presidential orders and without congressional authorization.

The question arises: By what authority did Roosevelt thus go to quasi-war in the North Atlantic? Looking back at the fiery debates of that ancient day, one is struck by the relative absence of constitutional argument. Isolationists denounced the Lend-Lease Act as an excessive delegation of legislative power to the President. But it was, after all, a statute duly passed by Congress after full debate. It was not a unilateral assumption of power by the President.

No isolationist had paid more attention to the Constitution than the formidable historian Charles A. Beard. Beard had made his name thirty years before with *An Economic Interpretation of the Constitution,* and in 1943 he published *The Republic,* a series of dialogues on the Constitution. But the two volumes of polemic against Roosevelt's foreign policy he wrote after the war turned on presidential violations of "covenants with the American people to keep this nation out of war"—covenants made in speeches and party platforms; not, except for scattered references in the epilogue to the second volume, on presidential violations of constitutional provisions and prohibitions.[29]

In Congress, isolationists tended to make substantive rather than constitutional arguments. Senator Robert A. Taft of Ohio was an exception. At one point he objected that the President had "no legal or constitutional right to send American troops to Iceland" without congressional authorization. Congressional acquiescence, Taft said, might "nullify for all time the constitutional authority distinctly reserved to Congress to declare war." But only one senator supported Taft's constitutional protest.[30] The failure to invoke the Constitution probably expressed a sense of futility about constitutional argumentation once the passage of the Lend-Lease Act had made Congress an accomplice in Roosevelt's policy.

The constitutional question remained in abeyance. Roosevelt acted as if his policies derived from the need to execute the congressional mandate embodied in the Lend-Lease Act, not from independent presidential or commander in chief power. Why then did he not seek explicit congressional authorization?

Unlike Lincoln, who could count on congressional ratification for his early unilateral measures, Roosevelt faced a bitterly divided Congress. He had to balance risks: the risk of arguably illegal actions that would get Lend-Lease goods to Britain against the risk, should he seek congressional authorization, of a defeat that would imply repudiation of the aid-to-Britain policy and might thereby, in Roosevelt's profound belief, mean the capitulation of Britain and deadly danger to the United States.

In April 1941, as British shipping losses grew, Henry L. Stimson, the Secretary of War, urged the President to request convoy authority from Congress. Roosevelt responded, as Stimson noted in his diary, "that it was too dangerous to ask the Congress for the power to convoy . . . If such a resolution were pressed now it would probably be defeated."[31] In May, Stimson handed Roosevelt a draft congressional resolution authorizing the use of force to protect the delivery of supplies to Britain. The President thanked him but again judged the time ill-chosen. In June, the President's Harvard classmate Grenville Clark, now an eminent lawyer, urged Roosevelt to ask Congress for a joint resolution approving measures necessary to assure delivery. Roosevelt replied in July that the time was not "quite right."[32] The renewal of selective service in August by a single vote in the House of Representatives would seem to vindicate the presidential assessment of the political odds.

Roosevelt's actions in the autumn of 1941, like Lincoln's in the spring of 1861, were, in a strict view, unconstitutional (though Lincoln's at least took place after war had begun). But, unlike later Presidents, Roosevelt did not seek to justify the commitment of American forces to combat by pleas of inherent constitutional power as President or as commander in chief. He thereby proposed no constitutional departures. Nor did he move behind a veil of secrecy. The debate between the isolationists and the interventionists was the most bitter in my lifetime. Roosevelt's major decisions were argued in the open and concluded in the open. With Hitler's cooperation, he brought the country along and kept it substantially united behind his policies.

He did not assert in the later imperial style that there was no need to consider Congress because the office of commander in chief gave him all the authority he needed. Jackson's opinion on the destroyer deal shows how undeveloped commander-in-chief theory was in those innocent days. In eighty-three press conferences in 1941 up to Pearl Harbor, Roosevelt never once alleged special powers in foreign affairs as commander in chief. When the title occurred in his speeches and messages, it generally signified only the narrow and traditional view of the commander in chief as the fellow who gave orders to the armed forces.

Pearl Harbor soon ended the policy debate. Thereafter Roosevelt, like Lincoln, had to cope with problems of internal security. Roosevelt had much the simpler task. It was easier to protect internal security in foreign war than in civil war. Moreover, civil liberties were themselves much more precisely defined and understood in 1941 than in 1861; and, as a result of the extension of the Bill of Rights by incorporation through the Fourteenth Amendment, civil liberties were in a much stronger constitutional position.

In 1940, while protesting his sympathy with Justice Holmes's condemnation of wiretapping in the Olmstead case, Roosevelt had granted the Attorney General qualified permission to wiretap "persons suspected of subversive activities against the United States."[33] Given the conviction Roosevelt shared with most Americans that a Nazi victory endangered the United States, he would presumably have been delinquent in his duty had he not ordered precautionary measures against Nazi espionage, sabotage, and "fifth column" penetration. Though we now know that the internal Nazi menace was even more exaggerated than the Copperhead menace had been, who could have been sure of that at the time? No more than Lincoln could Roosevelt foretell the outcome. Events now safely in the past were then in the perilous future.

Roosevelt, like Lincoln, broadened his apprehensions to include Americans honestly opposed to the war. By prodding the FBI to investigate isolationists and their organizations, he blurred the line between enemy agents and political opponents. Harking back to the Civil War, Roosevelt even called his isolationist adversaries Copperheads; and in the conspiracy-obsessed J. Edgar Hoover he found an equivalent of Lincoln's LaFayette Baker and Allan Pinkerton.

There was, however, little serious government follow-up of Roosevelt's prodding. His prods were evidently taken by his subordinates as expressions of passing irritation rather than of constant purpose. In 1941 Roosevelt appointed Francis Biddle, a former Holmes law clerk and a distinguished civil libertarian, as Attorney General. "The most important job an Attorney General can do in a time of emergency," Biddle said on assuming the office, "is to protect civil liberties. . . . Civil liberties

War Presidents. Two leaders who went beyond the Constitution in their defense of the United States and Liberty—as they understood them. Photograph of Lincoln by Anthony Berger, 1864, and detail of Roosevelt by anonymous, 1945. *Courtesy of James Mellon* and *the Franklin D. Roosevelt Library,* Hyde Park.

Roosevelt.

are the essence of the democracy we are pledged to protect."³⁴
Roosevelt kept Biddle on the job throughout the war despite
Biddle's repeated resistance to presidential importunings that
threatened the Bill of Rights.

Roosevelt's preoccupation with pro-Nazi activity increased
after Pearl Harbor. "He was not much interested in the theory
of sedition," Biddle later recalled, "or in the constitutional right
to criticize the government in wartime. He wanted this anti-
war talk stopped."³⁵ Biddle managed to avoid most presidential suggestions regarding the prosecution or suppression of
the press. But in time, Roosevelt's prods forced a reluctant Biddle
to approve the indictment of twenty-six pro-fascist Americans
under a dubious application of the law of criminal conspiracy.
A chaotic trial ended with the death of the judge, and the case
was dropped.

Biddle also opposed the most shameful abuse of power
within the United States during the war—the relocation and
detention of Americans of Japanese descent. Here Roosevelt
responded both to local pressure, including that of Attorney
General Earl Warren of California, and to the War Department, where such respected lawyers as Henry L. Stimson and
John J. McCloy argued for action. Congress ratified Roosevelt's executive order before it was put into effect, so the relocation did not represent a unilateral exercise of presidential
power. The Supreme Court upheld the program in the Hira-
bayashi and Korematsu cases, both decided, like the Prize cases,
in wartime.

The most vicious assaults on civil liberties in the Roosevelt
years resulted from private, not government, action—though
private action spurred on by the Supreme Court. The Gobitis
decision in 1940 upholding the compulsory salute and pledge
of allegiance to the flag led to persecutions of Jehovah's Wit-
nesses—who rejected flag worship as idolatry—mobs, arson,

and even a case of castration.[36] Then in 1943, despite the high patriotic fervor generated by the war, the Court reversed itself and declared the compulsory pledge and salute unconstitutional. "If there be any fixed star in our constitutional constellation," Robert H. Jackson, who was now an associate justice, wrote on behalf of the Court, "it is that no official, high or petty, can prescribe what shall be orthodox in politics, nationalism, religion, or other matters of opinion."[37] One would like to hope that these words still express the national view.

For all Roosevelt's moments of impatience and exasperation, his administration's performance on civil liberties during the Second World War—the Japanese-American case aside—was conspicuously better, if also easier to accomplish, than the Lincoln administration's performance during the Civil War. In 1945 the American Civil Liberties Union saluted "the extraordinary and unexpected record . . . in freedom of debate and dissent on all public issues and in the comparatively slight resort to war-time measures of control or repression of opinion."[38]

Most of Roosevelt's actions to protect national security—even the relocation of Japanese Americans—observed constitutional requirements of due process. His most conspicuous deviation from the Constitution during the war came in September 1942, when he told Congress that, if it did not repeal a particular provision in the Price Control Act within three weeks, he would refuse to execute it. "The President has the power, under the Constitution and under Congressional Acts," he declared, "to take measures necessary to avert a disaster which would interfere with the winning of the war."[39] Congress repealed the offending provision, averting a constitutional showdown.

The question lingers: By what authority did Roosevelt act? We are back again to Locke and emergency prerogative. Franklin Roosevelt had probably not looked at the *Second Treatise on Civil Government* since his student days at Harvard, if he had ever looked at it then. But the doctrine of emergency prerogative had endured because it expressed a real, if rare, necessity. Confronted by Hitler, Roosevelt supposed, as Jefferson and Lincoln had supposed in the crises of their presidencies, that the life of the nation was at stake and that this justified extreme measures, using "the sovereignty of Government," as Roosevelt said in 1941, "to save Government."[40] Like Jefferson and Lincoln, Roosevelt did not pretend to be exercising routine or inherent presidential power. Unlike Jefferson's case of the Burr conspiracy but like Lincoln's case of the Civil War, Roosevelt's case had substantial public backing, and the electorate (and therefore, as Mr. Dooley had predicted, the courts) sustained his use of emergency prerogative.

Roosevelt in 1941, like Lincoln in 1861, did what he did under what appeared to be a popular demand and a public necessity. Both Presidents took their actions in light of day and to the accompaniment of uninhibited political debate. They did what they thought they had to do to save the republic. They threw themselves in the end on the justice of the country and the rectitude of their motives. Whatever Lincoln and Roosevelt felt compelled to do under the pressure of crisis did not corrupt their essential commitment to constitutional ways and democratic processes.

National crisis, the law of self-preservation, the life of the republic at stake, might thus justify Lockean prerogative and the consequent aggrandizement of executive power. Lincoln and Roosevelt embraced the grim necessity. But, regarding executive aggrandizement as but a means to a greater end, the survival of liberty and law, of government by, for, and of the people, they used emergency power, on the whole, with discrimination and restraint. Nevertheless, they risked the creation of precedent. As the Supreme Court said soon after Appomattox, the nation had "no right to expect that it will always have wise and humane rulers. Wicked men, ambitious of power, with hatred of liberty and contempt of law, may fill the place once occupied by Washington and Lincoln."[41] How to assure the recession of executive power when the emergency passed?

Henry Adams, reflecting on the obliteration of the Constitution during the Civil War, observed that the Framers "did not presume to prescribe or limit the powers a nation might exercise if its existence were at stake. They knew that under such an emergency paper limitations must yield; but they still hoped that the lesson they had taught would sink so deep into the popular mind as to cause a reestablishment of the system after the emergency had passed." The test, Adams wrote in 1870, was now at hand. "If the Constitutional system restored itself, America was right."[42]

Lincoln and Roosevelt, seeing the war power as a means to a higher end, understood the need to restore the constitutional system and affirmed in the midst of the emergency that emergency prerogative must expire with the emergency. "The Executive power itself," said Lincoln, "would be greatly diminished by the cessation of actual war." "When the war is won," said Roosevelt, "the powers under which I act automatically revert to the people—to whom they belong."[43]

So indeed it happened, and the constitutional regime did reestablish itself. This was perhaps due less to renunciation by Presidents than to resistance by the people and resilience in the system. Lincoln had derided the notion that "the American people will, by means of military arrests during the rebellion lose the right of public discussion, the liberty of speech and the press, the law of evidence, trial by jury, and Habeas corpus throughout the indefinite peaceful future which I trust lies before them." He could not believe that, he said—once again the homely analogy—any more than he could believe that "a man could contract so strong an appetite for emetics during temporary illness as to persist in feeding upon them during the remainder of his healthful life."[44]

Once the crisis ended, the other two branches of government briskly reasserted themselves. The separation of powers sprang back to defiant life. A year after Lincoln's death, the Supreme Court held in *Ex parte Milligan* that the trial under martial law behind the lines of Lambdin P. Milligan, a venomously pro-slavery conspirator, violated the Constitution. Seward's elective kingship gave way in half a dozen years to a President at the bar of impeachment, followed by the period later famously characterized as one of "congressional government."

In the same fashion, the death of Roosevelt and the end of the Second World War were followed by a diminution of presidential power. A year after victory, Roosevelt's successor was so unpopular that voters said "To err is Truman" and elected a Republican Congress. The next year Congress gained posthumous revenge against the mighty wartime President by proposing the Twenty-second Amendment and thereby limiting all future Presidents to two terms in the White House.

The instinctive dialectic of politics thus offers a measure of insurance against the possibility that emergency prerogative might lead to post-emergency despotism. Yet the danger persists that power asserted during authentic emergencies may create precedents for transcendent executive power during emergencies that exist only in the hallucinations of the Oval Office and that remain invisible to most of the nation. The perennial question is: How to distinguish real crises threatening the life of the republic from bad dreams conjured up by paranoid Presidents spurred on by paranoid advisers? Necessity, as Milton said, is always "the tyrant's plea."

The experience of Lincoln and Roosevelt suggests, I believe, the standards that warrant presidential resort to emergency prerogative. The fundamental point is that emergency prerogative cannot be properly invoked on presidential say-so alone but only under stringent and persuasive conditions, both of threat and of accountability, with the burden of proof resting on the President.

Let me try to define these conditions. Here, I would submit, are the standards:

1. There must be a clear, present, and broadly perceived danger to the life of the nation and to the ideals for which the nation stands.

2. The President must define and explain to Congress and the people the nature and urgency of the threat.

3. The understanding of the emergency, the judgment that the life of the nation is truly at stake, must be broadly shared by Congress and the people.

4. Time must be of the essence; existing statutory authorizations must be inadequate; and waiting for normal legislative action must constitute an unacceptable risk.

5. The danger must be one that can be met in no other way than by presidential initiative beyond the laws and the Constitution.

6. Secrecy must be strictly confined to the tactical requirements of the emergency. Every question of basic policy must be open to national debate.

7. The President must report what he has done to Congress, which, along with the Supreme Court and ultimately the people, will serve as judge of his action.

8. None of the presidential actions can be directed against the domestic political process and rights.

These standards, I believe, sufficiently distinguished what Lincoln did in the spring of 1861 and Roosevelt did in the autumn of 1941 from what Jefferson did in 1807, from what Truman did in seizing the steel mills in 1952, from what Nixon did in his use of "national security" to justify illegal acts in 1972–73, from what Reagan did in the mid-1980s with regard to Iran and the contras, from what the Bush administration threatened to do (until Congress saved him by passing a resolution he deemed unnecessary) in going to war against Iraq in 1991.

Lincoln's policy after Sumter, Roosevelt's in the North Atlantic, at least in the eyes of most Americans at the time and of most scholars in retrospect, represented a necessity—but not a precedent. By declining to use claims of inherent and abiding presidential power to justify their actions, Lincoln and Roosevelt took care not to give lesser men precedents to be invoked against lesser dangers. These two Presidents remained faithful to the spirit, if not the letter, of the Constitution: acting on the spirit to save the letter.

"If the people ever let command of the war power fall into irresponsible and unscrupulous hands," Justice Jackson said in dissent in the Korematsu case, "the courts wield no power equal to its restraint. The chief restraint upon those who command the physical forces of the country, in the future as in the past, must be their responsibility to the political judgments of their contemporaries and to the moral judgments of history."[45]

7

War Opponent and War President

GABOR S. BORITT

Give me liberty, or give me death.

Patrick Henry, 1775

Here the truceless armies yet
Trample, rolled in blood and sweat;
They kill and kill and never die;
and I think that each is I.

A. E. Housman, "A Shropshire Lad"
XXVIII (1895)

On a blood red morning the sun
rises behind the form of a wakeful
sentry guarding the crest of Cemetery
Ridge, knees bent, as if in prayer.
Likely dead before the next sun.
Behind this bronze statue, behind
the roar of these cannon and this gun,
a bearded farmer rises in the mind

Of his own battle and calls in
the animals of the field. He
bends to each and hand to hand
milks the placid bovine arsenal
which lows contentedly, further than eye
    can see,
apart from the killing grounds of his
    blessed land.

James A. Miller, "Amishman at the New
World" (1986)

YEAR AFTER YEAR millions of Americans take the road to Gettysburg. In physical, moral, intellectual, or emotional terms many millions more all over the globe take in some ways similar pilgrimages. What are they looking for? The Gettysburg road leads to a peaceful, pastoral countryside— peaceful but for those few days in the summer heat of 1863— fields of hay and corn; fields of wheat; peaches, woods, cattle, sheep, graceful barns, neat homes, and working, peaceful people. The roadsides, the fields, and the woods are also spotted with countless monuments and markers forming what might be the most unusual outdoor art gallery in the world. And as the Gettysburg College folk, and the like, look at the milling pilgrims, the question burns into their inwards: What do the monuments honor? Whom they honor we know: the soldiers who fought here and died here. That is, in part, why Lincoln, the first and still chief pilgrim, came to Gettysburg. But what do the monuments honor? A number of answers can be given to this question, but among them we do not usually count, perhaps do not want to know, that when we honor the warrior, to a degree, we honor war itself. So it is "altogether fitting and proper" that the Gettysburg Address is a funeral oration.[1]

If Gettysburg is a place of monuments, so is in a different way its big neighbor, the capital city of Washington. There was put up in recent years, after much deliberation, a monument designed by an American woman of Oriental descent to those many sons and few daughters of this land who had died in Vietnam. This awesome granite wall stands between the Washington Monument and the Lincoln Memorial and close to the larger than life statue of a sitting, kindly Albert Einstein in sneakers, inviting children to sit in his lap, as my children did when we visited him. The black granite of the Vietnam monument, after a short time, was supplemented by a statue of three soldiers. The monument to the dead seems thus to have become a monument also to that majority of soldiers who had survived. Perhaps it would have been well to have also added—in this age of abstract art where most people are still mundanely literal-minded—the statue of a soldier-paraplegic in a wheelchair. One could make the argument that a wheelchair would have been "altogether fitting" next to the black of the Vietnam granite, in the shadow of the Greek temple honoring the man under whose leadership one and a half million casualties piled up in a civil war, and not far from the Egyptian obelisque honoring the father of Americans, whose casualties in the war at the birth of the nation were more modest, and with the statue of the gentle immigrant scientist nearby whose letter to Franklin Delano Roosevelt (soon to be another great war president) helped start the scientific project which reached its first success at Hiroshima.[2]

These allusions to the monuments of Gettysburg and Washington attempt to set a mood here and to suggest the emotional roots of the present inquiry. Historians, like other mortals, are children of their own times. My inheritance is that of the Vietnam generation—but with the shadow of the Second World War heavily upon it.

The intellectual and religious roots of this inquiry, however, go back millennia: Western civilization's concern about, and opposition to, war is ancient. The book of Isaiah—which Lincoln liked so much—and the book of Micah, too, prophesied a time when "they shall beat their swords into plowshares, and their spears into pruning hooks: nation shall not lift up sword against nation, neither shall they learn war any more." War in the Bible, at times, was punishment for the sins of an erring people. To teach a better way Jesus went up on the Mount and said "resist not evil: but whosoever shall smite thee on thy right cheek, turn to him the other also." His message was "love your enemies." The ancient Hebrew religion, of course, was also, in part, a great warrior religion, as was Christianity for more than a millennium, after Rome became a Christian empire under Constantine. Rome, as Greece before it, also had thinkers and leaders who justified the shedding of blood in the defense of liberty. Much blood was indeed shed in its defense—and much more otherwise.

St. Augustine of Hippo helped perfect the theological basis of the Christian Empire through the concept of the "just war," whereby the Christian soldier could smite the unjust enemy, but only with mercy in his heart. Augustine recognized the empire as the creator of "a bond of peace," but also grieved: "how many great wars, how much slaughter and bloodshed, have provided this unity!"[3]

A thousand years of wars ensued but Western society changed, and the Dutch pacifist Erasmus took up the anti-war spirit, as did the French monk Emeric Cruce, the German writer Grimmelshausen, and some others. But it was the eighteenth-century Enlightenment that shaped the modern liberal conscience. Sir Michael Howard, the British historian, defines that conscience as "pacific if not actually pacifist. It regards war as an unnecessary aberration. . . . On the other hand it accepts that wars may have to be fought, either to ensure the liberation of groups suffering under alien oppression, or to ensure the survival of those societies in which the liberal ethic has achieved dominance." Abraham Lincoln's attitude toward war, though he never expressed it systematically, is rooted in this tradition.[4]

Lincoln's first encounters with war were close to the mythical, a mixture of Biblical and American Revolutionary lore. The Bible had been touched on briefly, though it might be added that Lincoln knew the book and its stories, including its many war stories, in a way and to an extent that is rare in our own times. As for the Revolution, he was born thirty-four years after the first shots at Lexington Green, two decades after the establishment of the United States under the Constitution, a decade after the death of George Washington. In physical time the boy Lincoln was close to the Revolutionary War. In terms of understanding, however, by his day the War had already receded into fable. In contrast, what Americans liked to call the War of 1812, and which Lincoln tended to refer to as "the British War," coincided with his early childhood.

One of his first memories came from this war. One day as a little boy he went fishing and made a catch. On his way home he met a soldier coming back from the war. The boy gave his fish to the veteran—"having been always told at home that we must be good to soldiers." As if to explain, in later years his cousin Dennis Hanks remembered that the Lincoln home in Kentucky was a stop for the returning veterans of the war.[5]

There is no reason to doubt that such patriotic and in a

limited sense pro-war sentiments surrounded Lincoln's youth, and in later years on rare occasions he repeated, almost always in mild ways, such sentiments. But cousin Hanks remembered that it was Thomas Lincoln, the father who hailed the soldiers of 1812. Somehow, whether via the influence of the mother, Nancy Hanks, Lincoln's own internal imperatives, or a combination of further elements, another more dominant aspect emerged in him early. When many years later the historian Francis Parkman spoke of America's two greatest presidents, he expressed a preference for Washington. Lincoln had too much "womanly tenderness" in him, the scholar explained, thus repeating views expressed by men close to the Illinoisan. Whatever we may think of such sexual stereotyping today, what Parkman and the others meant was illustrated by Lincoln's attitude toward hunting.[6]

Hunting was a way of life on the frontier where he grew up, indeed the way to physical survival. But Lincoln refused to be very much a hunter. On the eve of his presidency he still remembered the traumatic experience of his childhood when "a few days before the completion of his eighth [*sic*] year, in the absence of his father, a flock of wild turkeys approached the new log-cabin [in which he lived], and A[braham]. with a rifle gun, standing inside, shot through a crack, and killed one of them. He has never since pulled a trigger on any larger game."[7]

Though the youthful Lincoln was involved in some pranks that seem cruel to late twentieth-century American eyes, his attitude toward hunting symbolized his rejection of the violence of the frontier in general. However much that frontier prized the use of physical prowess, the six-foot four wrestling champion of backwoods Illinois found that words, not violence, were the weapons acceptable to him. On the rare occasion he lapsed from his nonviolent standard, he remembered it ever after with tremendous shame, as in the case of his own near duel in 1841, which involved his courtship of Mary Todd. Such shame may have had various sources but we cannot ignore at its core a fundamental repugnance toward violence. It is not impossible that a seemingly apocryphal tale originated with a shamed Lincoln, who when challenged to the duel, chose for weapons cow dung at five paces. His actual choice of weapons, cavalry broadswords of the largest size, was only somewhat less ridiculous and in the end helped avert the duel because the tall, long-armed Lincoln could have been reached only with the greatest difficulty by his much shorter challenger. Since Lincoln could never be called a physical coward, the choice of weapons allows only one interpretation.[8]

Yet if Lincoln detested violence, at age twenty-three he was ready to defend his home. He thus acquired his sole military experience, as a citizen soldier for some weeks, during an Indian rising in 1832. His autobiography suggests that he volunteered because of economic necessity, to which we can add youthful patriotism and a belief in self-defense. But Lincoln saw no combat in what became known as the Black Hawk War. When he later referred to his "soldiering," he removed it as far as possible from a real war experience, speaking of it as consisting of "bloody struggles with the musquetoes [*sic*]" and "charges upon the wild onions."[9] As with so many of his public utterances, whether about war or other subjects, Lincoln had political purposes even in ridiculing himself as a warrior. But such purposes do not negate the significance of both anti-violent and anti-military sentiments he repeatedly expressed within the context of dominant culture that could look on violence with approval and prized military glory. Indeed, the most notable event of Lincoln's soldiering experience was his saving the life of an old Indian who, having come into the whites' camp with a safe conduct, was about to be lynched by the soldiers. The story has the stuff of legends, but it is nonetheless true.[10]

If Lincoln's war experience carried elements of both the ridiculous and the noble, with himself stressing the former, his memory also hid horror that he forever after could associate with war. Though Lincoln saw no action, he did see five dead, scalped men. As he came upon the men "the light of the morning sun was streaming upon them as they lay heads toward us on the ground. And every man had a round, red spot on the top of his head. . . . " The scalped men were "frightful" and "grotesque." Many years later he still remembered detail: "one man had on buckskin breeches." In those days Lincoln, too, wore buckskin breeches.[11]

As his political career progressed, for a number of reasons

Mighty Scourge of War. Photograph by Timothy O'Sullivan taken at Gettysburg on July 5, 1863. Until the pioneering work of Civil War photographers, Americans saw no realistic images of the war dead. Very few had actual war experiences like Lincoln's. *Courtesy of U.S. Army Military History Institute.*

Lincoln saw fit to cast votes against West Point—both as an Illinois State Representative and as a Congressman.[12] While the country was at peace, however, his pacific outlook found its most forceful expression in condemnation of the civil violence that plagued Andrew Jackson's America. Lincoln's Lyceum Address of 1838, now famous at least among scholars, identified violence as the grave threat against democracy. Indeed if he erred, it was not in his diagnosis of the ill, but in his belief that a religious adherence to the law would be a cure. This excessively optimistic liberal faith held that almost invariably peaceful solutions to the stressful problems of the nation could be found. The degree of blindness in this faith, combined with his feelings about violence—as well as political self-interest—helps explain why Lincoln, too, failed to heed what Robert V. Bruce calls "the shadow of a coming war," and insisted that there would be no civil war in America in his time.[13]

Perhaps Lincoln's optimism should have been tempered by the knowledge that in little more than four score years his countrymen had fought three wars. Indeed, the United States had been born in war and Lincoln had read histories of that war. His thoughts about the Revolution are noteworthy not for their praise of the sacrifice and courage of the founding generation, or the blessings these brought, but for showing few illusions about the means that brought the blessings. Fife and drum history so fashionable in his day and beyond held few charms for him.

Yet he did not share late twentieth-century qualms about such fare and, when pressed, could make political use of it. He could also slip into occasional patriotic oratory. "Every American, every lover of liberty," he said in 1838, should "swear by the blood of the Revolution. . . . " But they were to swear to uphold the laws and oppose violence. He praised George Washington, "the mightiest name of earth," but added "migh-

tiest in the cause of civil liberty." Lincoln supported benefits
for veterans, took pride in the fact that Americans had "permitted no hostile foot to desecrate [Washington's] resting place,"
but also understood "the powerful influence" the Revolution
"had upon the *passions* of the people as distinguished from their
judgement."[14]

Only once in a long career did Lincoln come demonstrably
close to speaking of the *military* glories of the Revolution, by
recalling a biography he had read "away back in my childhood"—Parson Mason Weems's *Life of Washington*. In 1861, in
Trenton, New Jersey, on his way to take up the presidency,
Lincoln mentioned Weems' book. "I remember all the ac-
counts there given of the battle fields and struggles for the
liberties of the country, and none fixed themselves upon my
imagination so deeply as the struggle here at Trenton, New
Jersey. The crossing of the river; the contest with the Hessians;
the great hardships endured at that time, all fixed themselves
on my memory. . . . " So at last in February in 1861, on a
revolutionary battlefield, Lincoln spoke directly of Washington
and his men as warriors. The president-elect was thus paying
his compliments to the local folks of Trenton, he was recalling
his own faraway childhood and, perhaps unconsciously, he was
getting ready for war.

But even in Trenton, and in Philadelphia's Independence
Hall, he emphasized a cause that, for him, was much bigger
than war: "I recollect thinking then, boy even though I was,
that there must have been something more than common that
those men struggled for." The war had been for more than
"National Independence . . . this Union, the Constitution,"
and the ordinary "liberties of the people." The war was for
"the original idea" of America "which gave promise that in due
time the weights should be lifted from the shoulders of all men,
and that *all* should have an equal chance."[15]

Lincoln also suggested, as early as 1838, that the Revolution, for all its achievements, inevitably employed the "basest principles" of human nature. He listed *"hate"* and lust for *"revenge"* as examples. Yet Lincoln thought that a war could also put to sleep such peacetime vices as "the jealousy, envy, and avarice, incident to our nature." Since baseness was present in peacetime, too, he thus almost ended praising war in the abstract.[16] By the early forties, however, Lincoln's words left no ambiguity. Even as he gave his broadest and most thoughtful praise to the Revolution, the holiest of the American holies, speaking of it as the event that created unprecedented political freedom, proved man's capacity to "govern himself," and planted the germ destined to grow into "the universal liberty of mankind," he also identified the terrible nature of that and all war. "It breathed forth famine, swam in blood and rode on fire; and long, long after, the orphan's cry, and the widow's wail, continued to break the sad silence that ensued."[17]

Lincoln's strongest words, and stand, against war, however, emerged while he was serving his only term in Congress, during the Mexican War. This was the first war of his adult life and Lincoln became its leading Illinois opponent, overcoming his initial, conventionally patriotic reaction to the outbreak of hostilities. By this time he had developed his own robust version of a Whig outlook that promised to build up the nation internally and open the door of advancement to the many. He had read liberal economists who opposed war from the perspective of political economy. Politics of course also intertwined with Lincoln's moral revulsion to the Mexican War, as opposition to it became largely a party matter. Yet it is difficult to miss the fundamental anti-war meaning of his 1848 stand. He denounced the president of the United States, James K. Polk, for provoking the conflict: "The blood of this war, like the blood of Abel, is crying to Heaven against him." Lincoln made no apologies for attacking the commander in chief, for throughout history rulers "had always been . . . impoverishing their people in wars, pretending . . . that the good of the people was the object." This, he argued, was "the most oppressive of all Kingly oppressions." "Military glory," Lincoln defined as "that attractive rainbow, that rises in showers of blood—that serpent's eye, that charms to destroy." To his young friend and law partner, Billy Herndon, who wrote from Illinois complaining vehemently about Lincoln's stand, the Congressman replied with eloquence: "You are compelled to speak; and your only alternative is to tell the *truth* or tell a *lie*. I can not doubt which you would do." "Yours forever A. Lincoln."[18]

Most Whigs in Illinois appeared to follow him. One leader who did not, and who had earlier protested the War of 1812, explained: "No, by God, I opposed one war, and it ruined me, and hence forth I am for *War, Pestilence,* and *Famine*."[19] Lincoln based his passionately moral anti-war stand on the conviction that the United States had begun the war both unnecessarily and unconstitutionally. But he did not, perhaps dared not, vote against the prosecution of the war. With American troops deep in Mexico, like most anti-war congressmen, he dared not abandon the troops. He voted for supplies and veterans' benefits, recognized the bravery of the soldiers, and was even willing to accept some acquisition of land from Mexico. To "toast the *men*, but not the cause," was no easy stance to take, specially while attempting to uphold a moral standard. However, denunciation of the war aimed at moving the president toward peace. In this it was partly successful. In many ways an ordinary human and also a politician, Lincoln hoped to "distinguish" himself, too, with his Mexican War stand. In this he failed, and the Democrats in his state condemned bitterly the "corruption" and "treason" of this new "Benedict Arnold."[20]

The War Opponent. Collage created by Gabor Boritt from the earliest Lincoln photograph, probably taken during the Mexican War by N. C. Shepherd; statements about the war in Lincoln's hand; and comments about him and the war from the Illinois press. *Courtesy of James Mellon.*

One might suppose that this War helped Lincoln develop something of a view of history which, like the hopes of the pacifists of his time, held that as civilization progressed war might be eliminated. In 1859 he explained that "stranger" and "enemy" need not be synonymous though they have been mostly that since the beginning of history, "down to very recent times." If people would get to know each other a brighter future would be waiting. "To correct the evils," he thought, "great and small, which spring from want of sympathy, and from positive enmity, among *strangers,* as nations, or as individuals, is one of the highest functions of civilization."[21]

If the Mexican War influenced thoughtful people's understanding of history, it also produced war heros, many of whom hurried home to run for political office. Lincoln put a good face on it but did not much like it. It gave him a foretaste of what was to come. As president he would be threatened more directly than ever before with the use of military glory as a route to political station. In 1864 he beat off the challenge of both Republican general John C. Frémont and Democratic general George B. McClellan. Being by then, paradoxically, a war leader himself—and feeling both the war's justice and terror—he could not arraign his military opponents as military men *per se.* Yet long before 1864, Lincoln sensed that the use of military route to political power, even in its American variety through the electoral process, might not be the best for democracy.

He himself had been elected captain of volunteers in 1832, and that first election—not its military connotation—produced an unforgettable elation in him.[22] All the same, even at the start of his political life, and unlike myriad compatriots around him, he refused to capitalize on his military title and never ran for office as Captain Lincoln.

In some part his strong dislike of the use of military credits in politics stemmed from the practical problems of his Whig Party having had to contend with the fame of the hero of New Orleans, Andrew Jackson. Vis-à-vis that general, Lincoln indeed let slip uncharacteristic words of bitterness, the feelings of a politician whose party had been trounced again and again by candidates sheltered by "the ample military coat tail of Gen. Jackson." Lincoln's humor does not disguise the bile in his 1848 denunciation: "Yes Sir, that coat tail was used, not only for Gen. Jackson himself; but has been clung to, with the grip of death, by every democratic candidate since. . . . Like a horde of hungry ticks you have stuck to the tail of the Hermitage lion to the end of his life; and you are still sticking to it, and drawing a loathsome sustenance from it, after he is dead. A fellow once advertised that he had made a discovery by which he could make a new man out of an old one, and have enough of the stuff left to make a little yellow dog. Just such a discovery has Gen. Jackson's popularity been to you. You not only twice made President of him out of it, but have had enough of the stuff left, to make presidents of several comparatively small men since. . . . "[23]

A measure of Lincoln's bitterness may have come from the fact that in response to Jackson the Whigs themselves have been forced to resort to military men. In 1836 they ran several candidates, General William Henry Harrison, the hero of Tippecanoe, being one, though Lincoln voted for Hugh Lawson White of Tennessee. In 1840, Lincoln supported Harrison, the Whig nominee, in a "hurrah" campaign in which he mostly, but not entirely, ignored military credentials and tried to talk sense, above all economics. Eight years later, supporting yet another military candidate, Lincoln was still quite ready to say almost nothing in his campaign speeches about the martial fame of his party's general.

In 1848, Lincoln nonetheless supported Zachary Taylor because of the "General availability" of the hero of Buena Vista. Lincoln conceded in private that Taylor was not the best man for the job but a way to turn "the war thunder" against the Democrats who made the war, gloried in it, and stood to benefit from it. By supporting the war hero, Lincoln also tried to fend off the charges of treason that were leveled against him for his Mexican War stand. And by the time the Whigs nominated General Winfield Scott to oppose General Franklin Pierce in the 1852 presidential campaign, Lincoln had nearly stopped campaigning, though not without noting that the "attempt" to set up Pierce as a "great General, is simply ludicrous and laughable."[24]

Nothing demonstrates the ambivalence of Lincoln's support for the Whig generals better than his combining such support with the stinging condemnation of military coat tails. Forgetting Washington, not totally without justification, Lincoln thus suggested that the Whigs had been forced to have their generals because of what the Democrats had started in 1824 with Jackson. Lincoln harbored a bitter contempt for "fixing the public gaze upon the exceeding brightness of military glory." Such glory created from much split blood, he explained earlier, was a "serpent's eye." It corrupted his love: politics. Thus we have another dimension of Lincoln's anti-military feelings.[25]

Those feelings received one of their best outlets in Lincoln's humor. "By the way, Mr. Speaker," Lincoln said in the House of Representatives in 1848, "did you know I am a military hero? Yes sir; in the days of the Black Hawk war, I fought, bled, and came away. . . . It is quite certain I did not break my sword, for I had none to break; but I bent a musket pretty badly on one occasion." This was the speech in which Lincoln spoke of his brave "charges upon the wild onions" and his "good many bloody struggles with the musquetoes [*sic*]."[26]

If he ridiculed his own soldiering, he ridiculed the military pretensions of others with equal gusto. When Mexican War brigader Pierce was eulogized for his heroism, Lincoln picked out parts of the praise to emphasize their absurdity. Here is an example Lincoln cited from a "heroic" account: "As we approached the enemy's position, directly under his fire, we encountered a deep ditch, or rather a deep narrow, slimy canal, which had been previously used for the purpose of irrigation. It was no time to hesitate, so we both plunged in. The horse I happened to ride that day was a light active Mexican horse. This circumstance operated in my favor, and enabled me to extricate myself and horse after considerable difficulty. Pierce, on the contrary, was mounted on a large, heavy American horse, and man and horse both sank down and rolled over in the ditch. There I was compelled to leave him. . . . After struggling there, I cannot say how long, he extricated himself from his horse, and hurried on foot to join his command, & c."

"Now," asked Lincoln, "what right had a brigadier general, when approaching the enemy's position, and directly under his fire, to sink down and roll over in a deep slimy canal and struggle there before he got out, how long, another brigadier general cannot tell, when the whole of both their brigades got across that same 'slimy canal,' without any difficulty worth mentioning."[27]

Or there is Lincoln's burlesque of the militia. After a particularly pompous military funeral he appears to have remarked: "If General———had known how big a funeral he would have had, he would have died long ago."[28] Some saw the militia as the nation's chief protection: citizens bearing arms proudly, ready to defend freedom. Others could make jokes of it, and Lincoln among them once described with much relish a "fantastic" parade in his Illinois hometown: "We remember one of these parades . . . at the head of which, on horseback, figured our old friend Gordon Abrams with a pinewood sword, about nine feet long, and a paste-board cocked hat, from front to rear about the length of an ox yoke, and very much the shape of one turned bottom upwards. . . . "Flags they had too," Lincoln went on, and humorous signs, one of which he cited: "We'll fight till we run, and we'll run till we die." "That," Lincoln announced with satisfaction, "was the last militia muster" in his hometown.[29]

Lincoln's frequent association of humor with matters military was—as with much of his humor—more than the addition of a light touch to a serious subject. It rather seems he was trying to ridicule violence and war out of existence, at least out of his own reality. From his choice of the longest broadswords against his opponent in the duel he never fought, the trail goes to his admission (which prefaces his sketch of the Illinois military parade) that he knew "how the institution of chivalry was ridiculed out of existence by its fictitious votary Don Quixote."[30]

Stephen A. Douglas summed up matters during the Great Debates of 1858 revealingly, if less than accurately, even as he tried to make political capital out of his rival's old opposition to the Mexican War. Referring to the "war" being made upon him, Douglas, the incumbent Senator, proclaimed that "there is something really refreshing in the thought that Mr. Lincoln is in favor of prosecuting one war vigorously. It is the first war I ever knew him to be in favor of prosecuting. It is the first war that I ever knew him to believe to be just or constitutional."[31]

Abraham Lincoln: war opponent. Fair enough. But this brief sketch is incomplete. Lincoln had not merely opposed violence and war but had also a measure of acceptance, of all that he opposed, as facts of American life. Lincoln did, after all, support the Whig generals for the presidency and on rare occasion dutifully praised their heroism. The mockery of his own military experience and his description of the ludicrous military parade in his hometown were part of replies to Democratic criticism of Whig nominations of generals. Still, Lincoln's defense, made relatively easy by the way Democratic candidates themselves had been praised as war heroes, thus amounted to the caricaturing of matters military in general.

Once in a while Lincoln could use militarist-sounding language in his speeches. In 1840 he berated political opponents who did not fight with "powder and balls, because the smell of sulphur offends their nostrils." "We rose each fighting," he said in 1854, "grasping whatever he could first reach—a scythe—a pitchfork—a chopping axe, or a butcher's cleaver. . . ." But we should not make too much of such male language, for he was really talking politics, the workings of democracy, in short talking about what he saw as the best substitute for war and violence. During the Civil War he made the implications of his metaphors explicit when he explained that the war was to decide whether in a democracy there can be an appeal from the "ballot to the bullet."[32]

Yet there are other bits and pieces in Lincoln's record that we must not ignore. While campaigning for William Henry Harrison, he condemned Democratic candidate Martin Van Buren's "Janus-faced policy in relation to the war." That meant the War of 1812, which Lincoln appears to claim Van Buren both opposed and supported. The charge is particularly ironic in view of Lincoln's own stand on the Mexican War. But, as Mark E. Neely, Jr., has noted, if Lincoln needed any lessons concerning how dangerous war opposition could be to a politician, he supplied these lessons himself. Such opposition could be used against a presidential candidate, for example, twenty-eight years after the fact.[33]

Not surprisingly the two most pro-war statements of his antebellum career came from the period immediately after the

Mexican War. Both instances came in local addresses, in eulogies occasioned by the deaths of Zachary Taylor and Henry Clay.[34]

In the case of Clay, Lincoln saw fit to single out the Kentuckian's part in leading the United States into the War of 1812 and so standing up to "aggravated" "British aggressions." Lincoln even gave a brief imitation of a stirring speech by Clay. From 1852, indeed from any point in Lincoln's life, the historian is tempted to look ahead, to the Civil War. Lincoln's words are important chiefly because of that war. But in 1852 Lincoln was looking backward to the Mexican War, making amends for his own opposition to it before a people who increasingly accepted the war and assimilated it into their rose-hued nationalistic memory.[35]

Two years earlier, in a somewhat wooden eulogy for President Taylor, Lincoln had outdone his speech on Clay. While reviewing Taylor's career and extolling his virtues, Lincoln inevitably extolled military virtues, too. Even then, as did many others, he presented two sides of war, "victory and blood," "glory and grief," "pride and sorrow." But speaking of the brave dead he declared "I think of all these . . . as Americans, in whose proud fame, as an American, I too have a share." And by giving a rousing description of battle he went further than any other place in his ten volumes of collected works to present war in a positive light.[36]

He described the feelings of American soldiers in a besieged fort. The defenders heard the approach of a relief column and the sounds of battle. Their apprehension grew; the outcome meant life or death for them. Orated Lincoln: "And now the din of battle nears the fort and sweeps obliquely by; a gleam of hope flies through the half imprisoned few; they fly to the wall; every eye strained—it is—it is—the stars and stripes are still aloft! Anon the anxious brethren meet; and while hand strikes hand, the heavens are rent with a loud, long, glorious, gushing cry of victory! victory!! victory!!!'"

The leader of the relief column was Zachary Taylor. The war was with Mexico; the spot where the war began. Lincoln the war opponent indeed knew how to defend himself before the bar of the American public. War opponent?, he asked about himself, and political friends, in 1848. "The declaration . . . is true or false, accordingly as one may understand the term 'opposing the war.' "37

The man who took the oath of the presidential office in the spring of 1861 was not a pacifist but he was a pacific man. The above flourish notwithstanding, he abhorred violence. He prized the "Reign of Reason," the *"mind,* all conquering *mind."*38 He tried to hold on to anti-militaristic feelings, succeeding most of the time, and he harbored a resentment of military intrusion into political life. He felt the dread of war. Ballots and bullets he saw as hostile alternatives, war and violence as failures of democracy.

Yet such failures came. It was a paradox of the liberal faith, that the lover of peace had to be ready to fight wars to defend the survival of that faith. Lincoln's antebellum stance on the wars of his nation's history make clear, as indeed do his views on such wars as the European revolutions of 1848 or the liberation movements in Latin America,39 that for him war was an acceptable means. Sometimes war was the only means to overthrow alien oppression and attain national independence. So it had been, he suggested, in the American Revolution. War at times was also the only means to defend one's home from hostile forces. So it had been, Lincoln indicated, in 1812. He may have thus misjudged history, but he helped clarify his notions of what was a just war. Surely a fair share of the pacific aspect of his outlook stemmed from his widely held faith that "all the armies of Europe, Asia and Africa combined, with all the trea-

sure of the earth (our own excepted) in their military chest;
with a Buonaparte for a commander, could not by force, take
a drink from the Ohio, or make a track on the Blue Ridge, in
a trial of a thousand years."[40] The quotation also suggests that
for all of Lincoln's ridicule of his own soldiering, militarism,
and military pretensions, his thought encompassed authentic
heroes. Soldiers who fought in liberal wars deserved both honor
and special privileges. And war could play an important part
in history. The Revolutionary War, he seemed to believe, was
a most fundamental fact of American history.

Lincoln's contradictions then—if they were that—stemmed,
in part, from his liberal faith. In part they came from inside
himself. He rather recoiled from hunting, but could write a
poem about the excitement of a bear hunt—if, in the end, only
to mock human folly. He was, it is often said, a supreme real-
ist. He knew how to curb within himself extreme manifesta-
tions of tendencies that ran counter to the dominant cultural
values of his people. This, too, was part and parcel of his success in life. However unimportant his poem about the bear
hunt, and however empty his eulogizing of General Taylor,
each carried something of the authentic Lincoln. Together with
his liberal faith, they help explain why, when his time came,
"dreaded" as war was to him, he could "accept war" rather
than let the nation, and as he believed, liberty, perish.[41]

How the war opponent functioned as a war president, how
Lincoln's attitude toward war affected his leadership, and how
the Civil War altered his attitudes, needs detailed inquiry. But
a few cursory outlines can be suggested here. That much changed
in him we need not doubt. Almost immediately upon the com-
mencement of hostilities, his personal interest in peaceful dis-
coveries and inventions turned into a like interest in "the tools
of war," as Robert V. Bruce had shown long ago.[42] Lincoln
"accepted" a short, little war, it seems. He first called for 75,000

Man of Peace or War? In *The New York Illustrated News* Thomas Nast's view of Lincoln's 1861 inaugural address carried more conviction in its depiction of the new president as a god of war than as an angel of peace. *Courtesy of the Illinois State Historical Library.*

militia whose term of service was three months. We do not know whether he, or the nation, would have been able to "accept" in early 1861 the war they actually got, the greatest in American history which would claim one and a half million casualties. I think not. But Lincoln learned.

Though the Radicals around him forever claimed that he was too soft and too weak, it is fair to say that Lincoln grew into a great war leader who, to quote T. Harry Williams, "acted as commander in chief and frequently as general in chief."[43] He even contemplated taking to the field of battle. "Destroy the rebel army," he ordered his reluctant generals in the East. He made ever more terrible war, a people's war, a total war. After the horrifying battle of Fredericksburg, while the North mourned during the Christmas of 1862, one of his secretaries mused: "We lost fifty percent more men than did the enemy, [in fact he sharply understated the loss], and yet there is a sense in the awful arithmetic propounded by Mr. Lincoln. He says that if the same battle were to be fought over again, every day, through a week of days, with the same relative results, the army under Lee would be wiped out to its last man, the Army of the Potomac would still be a mighty host, the war would be over, the Confederacy gone. . . ."[44]

The message was kill and destroy, if necessary use a bullet that exploded inside the flesh. Long gone was the Lincoln of the first inaugural, who, in addition to expediency, showed shades of the classic liberal view of war as the horror that generally accomplishes nothing. In 1861, hoping to discourage civil war, he had told his disgruntled southern countrymen: "suppose you go to war, you cannot fight always; and when, after much loss on both sides, and no gain on either, you cease fighting, the identical old question[s] . . . are again upon you." But, to repeat, the president learned. This new war-making Lincoln demanded the overthrow of the social and political system of the South and adopted what James McPherson has called a "national strategy of unconditional surrender."[45]

The president was not allowed to forget that emancipation decrees, as David Brion Davis pointed out, have always been shrouded in violence or its threat.[46] Nevertheless, the ending of slavery, Lincoln argued, would end the only thing that could have caused war among Americans. Peace with freedom would be thus both just and lasting. Lincoln's ideas of *peace* would benefit from fresh study, too, but from the beginning of the war he maintained that "a great lesson of peace" should be "teaching all, the folly of being beginners of a war."[47]

So the exploding bullets. By 1864 Lincoln unleashed Sherman in Georgia, Sheridan in the Shenandoah Valley, and made war, some said, again barbaric. And this bloody war he won. What happened to the little boy who had shot the turkey in the Indiana wilderness and whose heart ached so? Did he remember in the White House that little boy when his own son, Tad, made a pet of the family's holiday turkey and then, when the time came for butchering, recoiled with horror and pleaded with his father for the life of the bird?[48]

The reprieve was given to the turkey by the father; and it was given to so many human beings who needed mercy that Lincoln's pardons grew legendary. But he suffered; his photographs show a face that changed in a few years from vigorous middle age to old. When novelist Michael Shaara shut his eyes, he could see Pickett's men, "Killer Angels," charging up Cemetery Ridge.[49] I cannot match his eloquence, but when I shut my eyes I see a descendant of Quakers, with so much of a Quaker heart in him, standing on the parapet at Fort Stevens in the summer of 1864, in the midst of Confederate General Jubal Early's raid on Washington. My professional colleagues must forgive me because the conclusion this passage leads to is not subject to proof. But I see this tall man, six-foot four, with

Toll of War. Lincoln in 1857 and 1865. Photographs by Alexander Hesler and Alexander Gardner. *Courtesy of James Mellon.*

a top hat on to exaggerate his height further, recognizable to all on both sides. He stands there, bullets whistle by, an officer falls close to him, but he just stands there, looking at the enemy—the man who in a few months, after Appomattox, upon hearing his wife use the word "enemy," would retort: "Enemies, never again must we repeat that word."[50] But now it is summer, 1864, and the tall so very weary man is standing on the parapet. Why? Again, numerous explanations can be suggested but I see a man standing there looking not at the Confederates, but God, saying silently: if I am wrong, God, strike me down.

Dereliction of duty, one might say; what business does the president have to expose himself thus? Indeed, a junior officer, later to become a justice of the Supreme Court, supposedly shouted at him, "Get down you fool"—and he did. But a day later Lincoln repeated the ritual.[51] War leader though Lincoln became, much of the war opponent remained in him. Patrick Henry's "Give me liberty, or give me death" was on people's lips, but in the White House the president seems to have told the story of the soldier of the War of 1812. During that war, it was fashionable for sweethearts of soldiers to make belts with mottoes sewn into them, and one young woman asked her man if he wanted his belt emblazoned with "Liberty or Death!" To which the soldier replied: that was a little strong, how about just "Liberty or be Crippled"?

"What do I want with war? I am no war man; I want peace more than any man in this country . . . ," Lincoln said in 1861, repeating the sentiment every year of his presidency.[52] What else would a politician say, one might interrupt, but it is difficult to avoid not merely his sorrow but also his anger at war. He could still describe, at least guerrilla war, in terms that reminded one of his comment that the American Revolution had given a stage to the "basest principles" of human nature.

Said Lincoln now: "Actual war coming, blood grows hot, and blood is spilled. . . . Deception breeds and thrives. . . . Every foul bird comes abroad, and every dirty reptile rises up."[53] Lincoln respected his fellow human beings and, as far as I know, in no other context did he deride them so bitterly. And since now he had to be the chief recruiting master of the Union armies, and chief officer of morale, the above words he confined to a private letter written in 1863. But even in public, a year later, he said this much: "War, at the best, is terrible, and this war of ours, in its magnitude and in its duration, is one of the most terrible. It has deranged business, totally in many localities, and partially in all localities. It has destroyed property, and ruined homes; it has produced a national debt and taxation unprecedented, at least in this country. It has carried mourning to almost every home, until it can almost be said that the 'heavens are hung in black.'" Yet Lincoln added to what was a denunciation of the war, all war (as he always added to his comments that "no man desires peace more ardently that I"), a harsh qualifier: "We accepted this war for an object, a worthy object, and the war will end when that object is attained. Under God, I hope it never will until that time."[54] And so Lincoln's life and work, his thought on war, is a poignant testimony to the liberal dilemma.

With the war, his presidency, and his life nearly over, Lincoln gave a brief, moral history of that war, as he saw it. In the second inaugural address it was again the war opponent speaking—mingled with a Biblical prophet still summoning war. In 1861, he said, "all dreaded" war, "all sought to avert it," yet "the war came." Each side "looked for an easier triumph, and a result less fundamental and astounding." "The prayers of both could not be answered; that of neither has been answered fully. The Almighty had His own purposes." Was the war punishment for the sin of slavery? he asked, and then went on. "Fondly do we hope—fervently do we pray—that this mighty scourge of war may speedily pass away. Yet, if God wills that it continue, until all the wealth piled by the bond-man's two hundred and fifty years of unrequited toil shall be sunk, and until every drop of blood drawn with the lash, shall be paid by another drawn with the sword, as was said three thousand years ago, so still it must be said 'the judgements of the Lord, are true and righteous altogether.' " [55]

"With malice toward none": St. Augustine would have approved. So have so many in the world since that windy spring day in the March of 1865. Should *we* do so, too, as the second millennium moves towards its end, and so approve of Lincoln's love of peace and fighting of war?

"But I have reached the end of my time, and have hardly come to the beginning of my task," Lord Acton once said on the occasion of a lecture, a lament that might be appropriated for these poor labors. [56] This chapter, like this book, began with Gettysburg, in our time, and it must end with our time. We think that history shines a light not only into the darkness of the past but also into the present and future. Some say the light is bright, some that it flickers. But when we look at war in our time, some also say that 1945 was the year one, as atomic scientists used to refer to it, or as a generation of Germans called it, *Jahr Null,* Year Zero. The past before that holds few lessons, if any.

The atom bomb, it is reasonably clear, has helped save our planet from a major conflagration for more than four decades because we did not dare use it. [57] It has served us rather like the choice of the largest of swords served the young Lincoln when he averted his duel. But what if we do use the weapon? Historian Richard Current asked about Lincoln at the dawn of the nuclear age: "If, in the 1860s, Yankee ingenuity had been equal to producing such a weapon would he have withheld the atom bomb? Or if, in the 1940s, he had been in Harry Truman's place, would he have spared Hiroshima?" What if Lincoln saw the bomb as the last weapon to defend liberty not merely "for today" but, to quote his words, "for a vast future also"?[58] Today we must multiply the ahistorical question by megatons even as we ask: Can this nation, or any nation, hope for a better, more decent leader? And so what usable lessons does Abraham Lincoln, the good man of good faith, war opponent and war president, have for us today? You, the reader, must decide.

A Scrawny Mars. Detail of John Tenniel's cartoon, "Vulcan in the Sulks," from *Punch*, 1865. *Courtesy of the* Punch *Library, London.*

Notes

Introduction

1. Roy P. Basler, ed., Marion Dolores Pratt and Lloyd A. Dunlap, asst. eds., *The Collected Works of Abraham Lincoln,* 9 vols. (New Brunswick, N.J., 1953–55), 7:514–15.

2. I here alter the meaning of words I borrow from Edmund Wilson, *Patriotic Gore: Studies in the Literature of the American Civil War* (New York, 1962), 130.

One: The Shadow of a Coming War—
Robert V. Bruce

1. Paul C. Nagel, *One Nation Indivisible: The Union in American Thought, 1776–1861* (New York, 1964), 260.

2. Edmund C. Burnett, ed., *Letters of Members of the Continental Congress,* 8 vols. (Washington, D.C., 1921–38), 3:509.

3. Nagel, *One Nation,* 261.

4. John C. Hamilton, ed., *The Federalist* (Philadelphia, 1888), 76–81.

5. Charles M. Wiltse, ed., *The Papers of Daniel Webster: Correspondence,* I (Hanover, N.H., 1974), 28.

6. *Debates and Proceedings in the Congress of the United States, 1789–1824* (42 vols., Washington, D.C., 1834–56), 7 Cong., 1 sess., 77.

7. William Maclay, *The Journal of William Maclay* (New York, 1965), 215.

8. Thomas P. Slaughter, *The Whiskey Rebellion* (New York, 1986), 5, 206, 212.

9. John C. Miller, *The Federalist Era, 1789–1801* (New York, 1960), 229, 239–42; John C. Miller, *Crisis in Freedom* (Boston, 1951), 169–75.

10. James M. Banner, Jr., *To the Hartford Convention* (New York, 1970), 93, 109–11.

11. Ibid., 115–18, 120; Nagel, *One Nation,* 238.

12. Banner, *To the Hartford Convention,* 338–40, 343–44.

13. James F. Hopkins, ed., *The Papers of Henry Clay,* II (Lexington, Ky., 1961), 780–81.

14. Glover Moore, *The Missouri Controversy, 1819–1821* (Lexington, Ky., 1953), 50, 93, 95.

15. Ibid., 207.

16. Charles Francis Adams, ed., *Memoirs of John Quincy Adams,* 12 vols. (Philadelphia, 1874–77), 5:210.

17. Moore, *Missouri Compromise,* 101, 105, 177.

18. Ibid., 172–73, 175, 218–20.

19. Adrienne Koch and William Peden, eds., *The Life and Selected Writings of Thomas Jefferson* (New York, 1944), 698.

20. *Niles's Weekly Register,* XXV, 100–101 (Oct. 11, 1828), 405 (Feb. 14, 1829).

21. Nagel, *One Nation,* 263.

22. Charles M. Wiltse, ed., *The Papers of Daniel Webster: Speeches and Formal Writings,* (Hanover, N.H., 1986), 1:347–48.

23. William W. Freehling, *Prelude to Civil War* (New York, 1966), 1–2.

24. Richard E. Ellis, *The Union at Risk: Jacksonian Democracy, States' Rights, and the Nullification Crisis* (New York, 1987), 48.

25. James D. Richardson, *A Compilation of the Messages and Papers of the Presidents,* 20 vols. (New York, 1917), 3:1218.

26. Freehling, *Prelude,* 2–3, 275.

27. Ibid., 206.

28. Laura A. White, *Robert Barnwell Rhett: Father of Secession* (New York, 1931), 24–25.

29. Freehling, *Prelude,* 206, 291–92.

30. Charles S. Sydnor, *The Development of Southern Sectionalism* (Baton Rouge, 1948), 219.

31. Merrill D. Peterson, *The Great Triumvirate: Webster, Clay, and Calhoun* (New York, 1987), 230.

32. Ellis, *Union,* 180–81.

33. Samuel F. Bemis, *John Quincy Adams and the Union* (New York, 1956), 454.

34. Glyndon G. Van Deusen, *William Henry Seward* (New York, 1967), 199.

35. Nagel, *One Nation,* 265–66.

36. *Congressional Globe,* 29 Cong., 2 sess., Senate, Feb. 19, 1847.

37. Avery O. Craven, *The Growth of Southern Nationalism, 1848–1861* (Baton Rouge, 1953), 41.

38. Avery Craven, *Edmund Ruffin, Southerner* (New York, 1932), 112.

39. White, *Robert Barnwell Rhett,* 118.

40. Holman Hamilton, *Prologue to Conflict: The Crisis and Compromise of 1850* (Lexington, Ky., 1964), 59; Peterson, *Great Triumvirate,* 458.

41. Wiltse, *Webster Papers: Speeches,* 2:547–48.

42. David M. Potter, *The Impending Crisis, 1848–1861* (New York, 1976), 118.

43. David M. Potter, *Lincoln and His Party in the Secession Crisis* (New Haven, 1942), 1.

44. Allan Nevins and Milton H. Thomas, eds., *The Diary of George Templeton Strong,* 4 vols. (New York, 1952), 2:283.

45. Potter, *Impending Crisis,* 263–64.

46. William E. Gienapp, *The Origins of the Republican Party, 1852–1856* (New York, 1987), 447.

47. Michael F. Holt, *Forging a Majority: The Formation of the Republican Party in Pittsburgh, 1848–1860* (New Haven, 1969), 199–200.

48. *Richmond* (Va.) *Examiner,* Dec. 3, 1859.

49. William T. Sherman, *Memoirs of General William T. Sherman* 2 vols. (New York, 1875), 2:381.

50. Rachel S. Thorndike, ed., *The Sherman Letters* (New York, 1894), 63, 80, 83; Ulysses S. Grant, *Personal Memoirs of U.S. Grant,* 2 vols. (New York, 1885), 214–15.

51. T. Harry Williams, *P. G. T. Beauregard: Napoleon in Gray* (Baton Rouge, 1955), 44; Douglas S. Freeman, *R. E. Lee: A Biography,* 4 vols. (New York, 1934–35), 1:414–15.

52. Jefferson Davis, *The Papers of Jefferson Davis* 6 vols. (Baton Rouge, 1971–89), 5:123, 147, 313; Dunbar Rowland, ed., *Jefferson Davis, Constitutionalist, His Letters, Papers and Speeches* 10 vols. (Jackson, Miss., 1923), 2:350–51.

53. Russell F. Weigley, *History of the United States Army* (New York, 1967), 190, 567; Davis, *Papers,* 6:240, 560, 598, 638, 669.

54. John Sherman, *John Sherman's Recollections of Forty Years* (Chicago, 1896), 169–72.

55. Williams, *P. G. T. Beauregard,* 34.

56. Roy F. Nichols, *The Disruption of American Democracy* (New York, 1948), 178–79.

57. Freehling, *Prelude,* 322–33.

58. John Hope Franklin, *The Militant South* (Cambridge, Mass., 1956), 228–30, 235–40.

59. Davis, *Papers,* 6:55n, 160–61, 228, 230, 277, 622–23, 626–28; Rowland, *Davis,* 3:359.

60. William L. Barney, *The Secessionist Impulse* (Princeton, 1974), 112–17, 218.

61. Davis, *Papers,* 6:370n; Hudson Strode, *Jefferson Davis: American Patriot* (New York, 1955), 363.

62. Sherman, *Memoirs,* 2:381.

63. Roy P. Basler, ed., Marion Dolores Pratt and Lloyd A. Dunlap, assist. eds., *The Collected Works of Abraham Lincoln,* 9 vols. (New Brunswick, N.J., 1953–55), 8:332–33.

64. Donald E. Reynolds, *Editors Make War: Southern Newspapers in the Secession Crisis* (Nashville, Tenn., 1971), 21.

65. Reynolds, *Editors,* 90; Dwight L. Dumond, *Southern Editorials*

on Secession (New York, 1931), 143–44, 190, 195, 199–201, 227, 312–14, 371.

66. Reynolds, *Editors,* 94, 121.

67. David H. Donald, *Charles Sumner and the Coming of the Civil War* (New York, 1965), 360; Howard C. Perkins, *Northern Editorials on Secession,* 2 vols. (New York, 1942), 1:41, 73, 84.

68. Kenneth M. Stampp, *And the War Came* (Baton Rouge, 1950), 7–9.

69. Potter, *Impending Crisis,* 423, 431–32; Potter, *Lincoln and His Party,* 47–49.

70. Reynolds, *Editors,* 20, 167, 174.

71. Ibid., 174; William L. Barney, *The Road to Secession* (New York, 1972), 161, 171, 175, 197; Dumond, *Southern Editorials,* 464–65; Barney, *Secessionist Impulse,* 234–35.

72. Steven A. Channing, *Crisis of Fear* (New York, 1970), 274–78.

73. Stampp, *And the War Came,* 14–15; Grant, *Memoirs,* 1:218; Potter, *Lincoln and His Party,* 77, 225, 245; Perkins, *Northern Editorials,* 1:358–59; Donald, *Sumner,* 367–68.

74. Perkins, *Northern Editorials,* 1:94, 212, 233, 366, 2:843–44; Stampp, *And the War Came,* 260; Thorndike, *Sherman Letters,* 86, 102, 108; Dumond, *Southern Editorials,* 446.

75. Robert W. Johannsen, *Stephen A. Douglas* (New York, 1973), 788–89, 818–19; Robert W. Johannsen, ed., *The Letters of Stephen A. Douglas* (Urbana, 1961), 504–5.

76. Stampp, *And the War Came,* 10–11, 25–28, 66, 68, 74–75, 261; Potter, *Lincoln and His Party,* 194n, 234n.

77. Davis, *Papers,* 4:20.

78. Strode, *Davis,* 363.

79. Davis, *Papers,* 4:35, 210–11, 213, 390–92; Rowland, *Davis,* 1:598–99; Strode, *Davis,* 381, 385, 387.

80. Basler, ed., *Collected Works of Lincoln,* 1:108–15.

81. Richard N. Current, *The Lincoln Nobody Knows* (New York, 1958). Gabor S. Boritt, "The Voyage to the Colony of Linconia," *Historian,* 37 (1975), 619–32, developed the idea of "avoidance" in a Lincoln context.

82. Basler, ed., *Collected Works of Lincoln,* 2:355, 364, 367, 372.

83. Ibid., 2:461; 3:308, 316.

84. Ibid., 4:95.

85. Potter, *Impending Crisis,* 482.

86. Potter, *Lincoln and His Party,* 247.

87. Basler, ed., *Collected Works of Lincoln,* 4:241, 244.

88. Ibid., 4:271.

89. Strode, *Davis,* 396–97.

90. Potter, *Lincoln and His Party,* 145, 147; Basler, ed., *Collected Works of Lincoln,* 4:172.

91. Theodore C. Pease and James G. Randall, eds., *The Diary of Orville Hickman Browning,* 2 vols. (Springfield, Ill., 1925, 1931), 1:453.

92. Basler, ed., *Collected Works of Lincoln,* 8:332.

Two: Lincoln and the Strategy of Unconditional Surrender— James M. McPherson

1. Earl Schenck Miers, ed., *Lincoln Day by Day: A Chronology 1809–1865,* 3 vols. (Washington, 1960), Vol. III: *1861–1865,* ed. by C. Percy Powell.

2. *David Homer Bates, Lincoln in the Telegraph Office* (New York, 1907).

3. Roy P. Basler, ed., Marion Dolores Pratt and Lloyd A. Dunlap, asst. eds., *The Collected Works of Abraham Lincoln,* 9 vols. (New Brunswick, N.J., 1953–55), 7:499.

4. John Henry Cramer, *Lincoln Under Enemy Fire* (New York, 1948). See also Chapter 7, n.50, infra.

5. Basler, ed., *Collected Works of Lincoln,* 7:476.

6. Mark E. Neely, Jr., *The Abraham Lincoln Encyclopedia* (New York, 1982); Gabor S. Boritt, ed., Norman O. Forness, assoc. ed., *The Historian's Lincoln: Pseudohistory, Psychohistory, and History* (Urbana, 1988) and *The Historian's Lincoln: Rebuttals. What the University Press Would Not Print* (Gettysburg, 1988); Don E. Fehrenbacher, *Lincoln in Text and Context: Collected Essays* (Stanford, 1987).

7. T. Harry Williams, *Lincoln and His Generals* (New York, 1952); Kenneth P. Williams, *Lincoln Finds a General,* 5 vols. (New York, 1949–59).

8. Carl von Clausewitz, *On War,* translated by Col. James J. Graham, 3 vols. (London 1911), 1:23, 3:121; Russell F. Weigley, *The American Way of War* (Bloomington, 1973), xvii; Alastair Buchan, *War in Modern Society: An Introduction* (New York, 1968), 81–82.

9. *War of the Rebellion: A Compilation of the Official Records of the Union and Confederate Armies* (Washington, 1880–1901), Series I, Vol. 34, Pt. 3, 332–33. Hereinafter cited as *O.R.*

10. T. Harry Williams, *Lincoln and His Generals,* 11.

11. Clausewitz, *On War,* 1:xxiii.

12. Basler, ed., *Collected Works of Lincoln,* 4:332.

13. Ibid., 437.

14. May 23, 1862.

15. *Personal Memoirs of U.S. Grant,* 2 vols. (New York, 1885), 1:368.

16. Basler, ed., *Collected Works of Lincoln,* 5:426, 6:257, 281; Howard K. Beale, ed., *Diary of Gideon Welles,* 3 vols. (New York, 1960), 1:370; Tyler Dennett, ed., *Lincoln and the Civil War in the Diaries and Letters of John Hay* (New York, 1939), 69.

17. *Memoirs of Grant,* I, 368–69; Burke Davis, *Sherman's March* (New York, 1980), 109.

18. *O.R.,* Ser. I, Vol. 17, Pt. 2, 150.

19. Basler, ed., *Collected Works of Lincoln,* 5:48–49.

20. Ibid., 344–46, 350.

21. Ibid., 4:506.

22. George B. McClellan, *McClellan's Own Story* (New York, 1886), 487–89.

23. Basler, ed., *Collected Works of Lincoln,* 5:144–46, 222–23.

24. Ibid., 317–19; *New York Tribune,* July 19, 1862.

25. Gideon Welles, "The History of Emancipation," *The Galaxy,* 14 (Dec. 1872), 872–73.

26. David Donald, ed., *Inside Lincoln's Cabinet: The Civil War Diaries of Salmon P. Chase* (New York, 1954), 149–52; Howard K. Beale, ed., *Diary of Gideon Welles,* 1:142–45; John G. Nicolay and John Hay, *Abraham Lincoln: A History,* 10 vols. (New York, 1890), 6:158–63.

27. The text of the preliminary and final proclamations is in Basler, ed., *Collected Works of Lincoln,* 5:433–36; 6:28–30.

28. Allan Nevins, *The War for the Union: War Becomes Revolution* (New York, 1960), 239; *O.R.*, Ser. I, Vol. 24, Pt. 3, 157; General Grenville Dodge quoted in Bruce Catton, *Grant Moves South* (Boston, 1960), 294.

29. Basler, ed., *Collected Works of Lincoln*, 6:149–50, 408–9. Lincoln was here repeating the words of General Grant (a prewar Democrat) who had written to him on August 23, 1863, in enthusiastic support of emancipation and black troops. Abraham Lincoln Papers, Library of Congress.

30. Edward Stanwood, *A History of the Presidency* (Boston, 1903), 301–2; Basler, ed., *Collected Works of Lincoln*, 7:23.

31. Dunbar Rowland, ed., *Jefferson Davis, Constitutionalist: His Letters, Papers, and Speeches*, 10 vols. (Jackson, Miss., 1923), 5:409; *O.R.*, Ser. II, Vol. 5, 797, 940–41.

32. Basler, ed., *Collected Works of Lincoln*, 7:435; Hudson Strode, *Jefferson Davis: Tragic Hero, 1864–1889* (New York, 1964), 77. For the abortive peace negotiations of 1864, see Edward C. Kirkland, *The Peacemakers of 1864* (New York, 1927), chs. 2–3.

33. Basler, ed., *Collected Works of Lincoln*, 8:151.

34. Ibid., 7:499–501, 506–7.

35. Ibid., 501, 517; Nicolay and Hay, *Abraham Lincoln*, 9:221.

36. Basler, ed., *Collected Works of Lincoln*, 8:151.

37. Ibid., 279; Strode, *Jefferson Davis*, 140–41.

38. Francis B. Carpenter, *Six Months at the White House with Abraham Lincoln* (New York, 1866), 77.

39. Basler, ed., *Collected Works of Lincoln*, 8:333.

Three: The Emancipation Moment—
David Brion Davis

1. Martin Luther King, Jr., "I Have a Dream," in Flip Schulke, ed., *Martin Luther King, Jr.: A Documentary . . . Montgomery to Memphis, with an introduction by Coretta Scott King* (New York, 1976), 218.

2. Ralph Wardlaw, *The Jubilee: A Sermon Preached in West George-*

Street Chapel, Glasgow, on Friday, August 1st, 1834, the Memorable Day of Negro Emancipation in the British Colonies (Glasgow, 1834), 13, 16–17, 20.

3. Ralph Waldo Emerson, "Address Delivered in Concord on the Anniversary of the Emancipation of the Negroes in the British West Indies, August 1, 1844," in *Complete Works of Emerson,* 14 vols. (New York, 1903–04), 11:99, 115–16, 135.

4. *Addresse de la Société des Amis des Noirs, à l'assemblée nationale, à toutes les villes de commerce, à toutes les manufactures, aux colonies, à toutes les sociétés des amis de la constitution* (2nd ed., Paris, 1791), 107–8.

5. Thomas Fowell Buxton to Mr. East, Oct. 15, 1832, Buxton Papers, III, 31–32, Rhodes House, Oxford.

6. James Stephen, *The Slavery of the British West India Colonies Delineated* . . . 2 vols. (London, 1824–30), 2:401–2.

7. James Stephen, Jr., Draft of a circular dispatch, Jan. 1833, Grey Papers, University of Durham, Department of Palaeography and Diplomatic.

8. James Stephen, Jr., Two commentaries on Lord Howick's plan, July 6, 1832, Grey Papers.

9. League of Nations, *Publications,* A. 19. 1925 VI, II.

10. Orlando Patterson, *Slavery and Social Death: A Comparative Study* (Cambridge, Mass., 1982), 209–96.

11. Louis, chevalier de Jaucourt, "Traite des nègres," *Encyclopédie, ou dictionnaire raisonné des sciences, des arts et des métiers,* XVI (1765), 532.

12. John Kenrick, *Horrors of Slavery* (Cambridge, Mass., 1817), 58–59; Andrew Thomson, *Immediate Emancipation* (Manchester, 1832), 4, 11, 24.

13. Roy P. Basler, ed., Marion Dolores Pratt and Lloyd A. Dunlap, asst. eds., *The Collected Works of Abraham Lincoln,* 9 vols. (New Brunswick, N.J., 1953–55), 5:433–36, 530–37.

14. Belle Becker Sideman and Lillian Friedman, eds., *Europe Looks at the Civil War* (New York, 1960), 190–91.

*Four: One Among Many: The United States
and National Unification—
Carl N. Degler*

1. For the story on Garibaldi and the Lincoln Administration, see Howard R. Marraro, "Lincoln's Offer of a Command to Garibaldi: Further Light on a Disputed Point of History," *Journal of the Illinois State Historical Society,* 36 (1943), 237–70.

2. Gladstone is quoted in John Gooch, *The Unification of Italy* (London, 1986), 1.

3. Denis Mack Smith, ed., *The Making of Italy, 1776–1870* (New York, 1968), 10–11.

4. See Chapter 1, supra.

5. Paul C. Nagel, *One Nation Indivisible. The Union in American Thought* (New York, 1964), chap. 1.

6. Quoted in Nagel, *One Nation Indivisible,* 38,62.

7. David M. Potter, *The Impending Crisis 1848–1861* (Completed and edited by Don E. Fehrenbacher, New York, 1978), 469.

8. See, for example, John McCardell, *The Idea of a Southern Nation: Southern Nationalists and Southern Nationalism, 1830–1860* (New York, 1979), and Drew Gilpin Faust, *The Creation of Confederate Nationalism* (Baton Rouge, 1988).

9. For a recent delineation of differences between North and South that does not contrast bourgeois and pre-bourgeois, see Randall C. Jimerson, *The Private Civil War: Popular Thought During the Sectional Conflict* (Baton Rouge, 1988).

10. Faust, *The Creation of Confederate Nationalism,* 4–7; James J. Sheehan, *German History, 1770–1866* (Oxford, 1989), 868–69. See also Benedict Anderson, *Imagined Communities: Reflections on the Origin and Spread of Nationalism* (London, 1983), and E. J. Hobsbawm, *Nations and Nationalism Since 1780. Programme, Myth, Reality* (Cambridge, 1990), 13–14, where he describes nation and nationalism as "exercises in social engineering which are often deliberate and always innovative."

Don Fehrenbacher draws a suggestive comparison between Southern nationalism in 1861 and American nationalism in 1776 in his

Constitutions and Constitutionalism in the Slaveholding South (Mercer University Lamar Memorial Lecture, No. 31, Athens, Ga., 1989), 59–60.

11. Quoted in James M. McPherson, "Antebellum Southern Exceptionalism: A New Look at an Old Question," *Civil War History,* 29 (1983), 233. This article splendidly develops the argument that it was the North, not the South, which changed over the course of the nineteenth century.

12. Allan Nevins, *Ordeal of the Union,* 4 vols. (New York: Scribners, 1947–50), 2:533–54. Emphasis in original. Kenneth Stampp, on the other hand, has strongly rejected the idea of the emergence of separate peoples: "Except for the institution of slavery," he wrote in 1980, "the South had little to give it a clear national identity. It had no natural frontiers, its white population came from the same stocks as the Northern population; its political traditions and religious beliefs were not significantly different from those of the North; it had no history of its own; and the notion of a distinct Southern culture was largely a figment of the romantic imagination of a handful of intellectuals and proslavery propagandists." Kenneth M. Stampp, *The Imperiled Union. Essays on the Background of the Civil War* (New York, 1980), 255–56. But as Ernest Renan shrewdly pointed out, "To forget and—I will venture to say—to get one's history wrong, are essential factors in the making of a Nation; and thus the advances of historical studies is often a danger to nationality." Ernest Renan, "What Is a Nation?," in Alfred Zimmern, ed., *Modern Political Doctrines* (London, 1939), 190.

13. See his provocative essay "The Southern Road to Appomattox," in his *Imperiled Union.*

14. Gladstone quoted in Belle Becker Sideman and Lillian Friedman, eds., *Europe Looks at the Civil War* (New York: 1969), 186; Acton's letter in J. Rufus Fears, ed., *Selected Writings of Lord Acton,* 3 vols. (Indianapolis, 1985–88), 1:363.

15. Erich Angermann, *Challenges of Ambiguity; Doing Comparative History;* German Historical Institute Annual Lecture Series, No. 4 (New York, 1991).

16. Renan, "What Is a Nation?," 190.

17. Rudolf Jhering the German jurist described Bismarck's effort in 1866 to revise the old German Confederation as "Germans armed against Germans, a civil war. . . ." Quoted in Theodore S. Hamerow, *The Social Foundations of German Unification, 1858–1871: Struggles and Accomplishments* (Princeton, 1972), 261. See also Sheehan, *German History*, 899, and Hans-Ulrich Wehler, *The German Empire, 1871–1918* (trans. Kim Taylor, Dover, N.H., 1985), 25 where they specifically write of the war between Austria and Prussia as a German civil war.

18. The most conspicuous example was Bismarck's so-called "closing of the gap" in the Prussian Constitution in 1862, an interpretation that effectively ended parliamentary government in Prussia between 1862 and 1866.

19. On Bismarck's alterations in laws affecting the modernization of the economy, see Hamerow, *Social Foundations,* 338–45.

20. David M. Potter, "Civil War," in C. Vann Woodward, ed., *The Comparative Approach to American History* (New York, 1968), 143.

21. Quoted in James M. McPherson, *Battle Cry of Freedom. The Civil War Era* (New York, 1988), 253.

22. Ibid., 859. See also McPherson, *Abraham Lincoln and the Second American Revolution* (New York, 1991).

23. For a vigorous pursuit of the argument that Lincoln skillfully maneuvered the South into firing the first shot, see John Shipley Tilley, *Lincoln Takes Command* (Chapel Hill, 1941), chap. xv, in which Tilley quotes Lincoln as taking "no small consolation" from the failure to resupply Sumter because the effort was "justified by the result," that is, the response of the North. See also Ludwell H. Johnson, *Division and Reunion: America 1848–1877* (New York, 1978), 78–79, for other evidence pointing in the same direction. Richard N. Current, *Lincoln and the First Shot* (New York, 1963), provides a very different interpretation.

24. James G. Randall, *Constitutional Problems Under Lincoln* (revised ed., Gloucester, Mass., 1963), 513–14. At one point in his book, in discussing the constitutionality of the blockade, which Lincoln initiated, Randall came close to identifying Bismarck's and Lincoln's repudiation of constitutional limitations. If the Supreme Court in the

Prize Cases had found that Lincoln exceeded his powers in proclaiming the blockade and expanding the army and navy, Randall wrote, then the Court "would have seemed to legitimize a dictatorship analagous to that of Bismarck from 1862–1866" (ibid., 57). For different views see Chapter 6, supra, and also Mark E. Neely, Jr., *The Fate of Liberty: Abraham Lincolns and Civil Liberties* (New York, 1991).

25. Quoted in Randall, *Constitutional Problems,* 378.

26. Wendell Phillips, *Speeches, Lectures and Letters* (Boston, 1864), 350. Better known than Phillips's statement of rejection of a Union held together by force is Horace Greeley's "We hope never to live in a republic whereof one section is pinned to the residue by bayonets." Although David M. Potter many years ago sought to show that Greeley never meant to sanction peaceable disunion, his exegesis of Greeley's text failed to recognize that the above quoted statement referred to the nature of the Union and not to avoiding war, as Potter contended. See David M. Potter, "Horace Greeley and Peaceable Secession," *Journal of Southern History,* 7 (1941), 156.

27. John Quincy Adams, *Jubilee of the Constitution,* (New York, 1839), 68–69.

28. Charles Sumner, *Works,* 15 vols. (Boston, 1877), 12:187–249.

29. Ibid., 246, 193.

30. Ibid., 204.

31. The similarity between the conservative (and romantic) character of the Confederacy and that of the *Sonderbund* is strikingly evident in the remarks of Franz von Elgger, a leader of the secessionists, after the defeat. "I fought and suffered not as a standard bearer of a party but as a citizen of Lucerne . . . as a Swiss for the ancient Confederation, for the heritage of our fathers, for true liberty, for the independence of our fatherland. . . . Fate has condemned me to survive the day of shame, yet I stand upright with a clear conscience and I throw my broken sword on the coffin of the old Switzerland." Quoted in Hans Kohn, *Nationalism and Liberty. The Swiss Example* (London, 1956), 101. Jefferson Davis and Alexander Stephens after the American war expressed almost identical ideas, but they wrote whole books to say it.

32. Kohn, *Nationalism and Liberty,* 102. Kohn emphasizes the deliberateness with which the Swiss drew upon the United States Constitution by pointing out that the idea of a bicameral legislature had no precedence in Swiss history.

33. Unlike the situation in the United States in the aftermath of its war, the cantons of the *Sonderbund* were required to pay the full cost of the war they had just lost. James Murray Luck, *A History of Switzerland* (Palo Alto, 1985), 361–62, 367.

34. Robin W. Winks, *Canada and the United States: The Civil War Years* (Baltimore, 1960), 376.

35. Ibid., 380.

36. A similar argument has appeared in print by an English historian of Germany writing in 1989. "German national consciousness no longer exists even in the sense in which it did, within limits, between 1871 and 1945," wrote Richard Evans. "There is no fundamental reason why a linguistic or cultural group such as the Germans should need to be united under a single state." Richard J. Evans, *In Hitler's Shadow. West German Historians and the Attempt to Escape from the Nazi Past* (New York, 1989), 102.

37. Quoted in Stampp, *Imperiled Union,* 36.

Five: One Alone? The United States and National Self-determination— Kenneth M. Stampp

1. Dexter Perkins, *A History of the Monroe Doctrine* (Boston, 1963), 28–64; George Dangerfield, *The Awakening of American Nationalism* (New York, 1965), 163–66. For an excellent account of American support of the principle of self-determination in the nineteenth century, see Betty Miller Unterberger, "National Self-determination," in Alexander DeConde, ed., *Encyclopedia of American Foreign Policy,* 3 vols. (New York, 1978) 2:635–38. For broader critical analyses of the principle, see Alfred Cobban, *The National State and National-Self determination* (New York, 1969), and Eric J. Hobsbawm, *Nations and Nationalism Since 1780* (Cambridge, 1990), esp. 5–13, 131–62.

2. Glyndon G. Van Deusen, *William Henry Seward* (New York,

1967), 139–40; Robert W. Johannsen, *Stephen A. Douglas* (New York, 1973), 329–31; Irving H. Bartlett, *Daniel Webster* (New York, 1978), 261–63.

3. Samuel Flagg Bemis, *A Diplomatic History of the United States* (Rev. ed., New York, 1942), 624–27.

4. Roy P. Basler, ed., Marion Dolores Pratt and Lloyd A. Dunlap, assist. eds., *The Collected Works of Abraham Lincoln,* 9 vols. (New Brunswick, N.J., 1953–55), 4:264–65.

5. For the evolution of the idea of a perpetual Union see Paul C. Nagel, *One Nation Indivisible: The Union in American Thought, 1776–1861* (New York, 1964), and Kenneth M. Stampp, "The Concept of a Perpetual Union," *Journal of American History,* 65 (June 1978), 5–33.

6. Basler, ed., *Collected Works of Lincoln,* 4:264.

7. James E. Cooke, ed., *The Federalist* (Cleveland, 1961), 254, 313; *Letters and Other Writings of James Madison,* 4 vols. (Philadelphia, 1867), 1:344; Stampp, "The Concept of a Perpetual Union," 13–20.

8. Nagel, *One Nation Indivisible,* 13–31. Quote from John Randolph on page 19.

9. James D. Richardson, ed., *A Compilation of the Messages and Papers of the Presidents,* 10 vols. (Washington, 1899), 1:213–14.

10. Quoted in Merrill D. Peterson, *Thomas Jefferson and the New Nation: A Biography* (New York, 1970), 772.

11. Quoted in Nagel, *One Nation Indivisible,* 19.

12. Stampp, "The Concept of a Perpetual Union," 28.

13. Richardson, ed., *Messages and Papers of the Presidents,* 2:640–56; Stampp, "The Concept of a Perpetual Union," 28–33.

14. Calvin Colton, ed., The *Private Correspondence of Henry Clay* (Cincinnati, 1856), 313. For the evolution of the state-rights and secessionist arguments, see Jesse T. Carpenter, *The South as a Conscious Minority, 1789–1861: A Study in Political Thought* (New York, 1930).

15. Richardson, ed., *Messages and Papers of the Presidents,* 2:755–56.

16. Kenneth M. Stampp, *America in 1857: A Nation on the Brink* (New York, 1990), 104–8.

17. Frank Moore, ed., *The Rebellion Record,* 12 vols. (New York, 1861–68), 1:3–4.

18. David M. Potter, *The South and the Sectional Conflict* (New

York, 1968), 68–69; Richard N. Current, *Northernizing the South* (Athens, Ga., 1983), 14; Kenneth M. Stampp, *The Imperiled Union: Essays on the Background of the Civil War* (New York, 1980), 255–56. For the argument among historians about whether Southerners were becoming a "separate people," see Chapter 4 n7–12, supra.

19. Moore, ed., *Rebellion Record,* 1:45.

20. Basler, ed., *Collected Works of Lincoln,* 4:265.

21. Philadelphia *Press,* April 13, 1861.

22. New York *Daily Tribune,* November 2, 9, 16, 26, December 3, 8, 24, 1860; David M. Potter, *Lincoln and His Party in the Secession Crisis* (New Haven 1942), 51–57.

23. Boston *Daily Advertiser,* November 12, 1860.

24. Potter, *Lincoln and His Party in the Secession Crisis,* 219–79.

25. The compromise movement is treated at length in Mary Scrugham, *The Peaceable Americans* (New York, 1921); Dwight L. Dumond, *The Secession Movement, 1860–1861* (New York, 1931); Potter, *Lincoln and His Party in the Secession Crisis;* and Kenneth M. Stampp, *And the War Came: The North and the Secession Crisis, 1860–1861* (Baton Rouge, 1950).

26. New York *Times,* December 1, 10, 1860.

27. Basler, ed., *Collected Works of Lincoln,* 4:264, 270.

28. Ibid., 1:438, 2:115.

29. Ibid., 4:434n.

30. Ibid., 4:426. See also Thomas J. Pressly, "Bullets and Ballots: Lincoln and the 'Right of Revolution,'" *American Historical Review,* 67 (1962), 647–62.

31. Basler, ed., *Collected Works of Lincoln,* 4:149–51, 175–76; Stampp, *And the War Came,* 185–87.

32. Basler, ed., *Collected Works of Abraham Lincoln,* 4:138–40.

33. Ibid., 4:154.

34. Ibid., 4:158–59, 162, 164.

35. Theodore C. Pease and James G. Randall, eds., *The Diary of Orville Hickman Browning,* 2 vols., (Springfield, Ill., 1927), 1:453–54.

36. Basler, ed., *Collected Works of Lincoln,* 4:195, 237, 240–41, 266, 171; Stampp, *And the War Came,* 189–97.

37. Springfield (Mass.) *Daily Republican,* March 6, 1861.

38. Basler, ed., *Collected Works of Lincoln,* 4:280, 292–93.

39. Ibid., 350–51. For two views of the Sumter crisis see Potter, *Lincoln and His Party in the Secession Crisis,* 336–75; Richard N. Current, *Lincoln and the First Shot* (Philadelphia and New York, 1963); Stampp, *And the War Came,* 262–86.

40. Moore, ed., *Rebellion Record,* 1:166–75; James M. McPherson, *Abraham Lincoln and the Second American Revolution* (New York, 1990), 27–29.

41. *Congressional Globe,* 37 Cong., 1 Sess., 222–23, 265; Basler, ed., *Collected Works of Lincoln,* 4:263, 439.

42. J. W. Bliss to Charles Sumner, December 19, 1860; January 1, 1861, Sumner Papers, Harvard University; Neal Dow to John A. Andrew, January 19, 1861, Andrew Papers, Massachusetts Historical Society; Wendell Phillips, *Speeches, Lectures, and Letters* (Boston, 1863), 396; Stampp, *And the War Came,* 247–52, 292–93.

43. James M. McPherson, *Battle Cry of Freedom* (New York, 1988), 494–500.

44. Basler, ed., *Collected Works of Lincoln,* 5:144–46, 529–34.

45. David Donald, *Charles Sumner and the Rights of Man* (New York, 1970), 58–64; Philip S. Foner, ed., *The Life and Writings of Frederick Douglass,* 5 vols. (New York, 1950–75), 3:256; Walter M. Merrill, ed., *The Letters of William Lloyd Garrison, Volume V: Let the Oppressed Go Free, 1861–1867* (Cambridge, Mass., 1979), 112–13; New York *Daily Tribune,* August 19, 1862.

46. Basler, ed., *Collected Works of Lincoln,* 5:388–89.

47. Ibid., 5:433–36, 6:28–30: Richard Hofstadter, *The American Political Tradition* (New York, 1948), 131.

48. Basler, ed., *Collected Works of Lincoln,* 6:48–49.

49. Hofstadter, *American Political Tradition,* 132–34.

50. Basler, ed., *Collected Works of Lincoln,* 6:319–20.

51. Ibid., 7:23.

52. Ibid., 6:358, 365, 408–11; 7:51, 500; 8:152.

53. Ibid., 7:435.

54. Ibid., 8:254–55.

55. Merrill, ed., *Letters of William Lloyd Garrison,* 5:258.

56. Basler, ed., *Collected Works of Lincoln,* 8:332–33.

57. Hobsbawm, *Nations and Nationalism Since 1780,* 132–33, notes "the utter impracticability of the Wilsonian principle to make state frontiers coincide with the frontiers of nationality and language." The new states after 1918 "were quite as multinational as the old 'prisons of nations' they replaced."

58. Karl E. Meyer, "Woodrow Wilson's Dynamite," *New York Times,* August 14, 1991; Arthur Schlesinger, Jr., "Self-determination: Yes, but . . ." *Wall Street Journal,* September 27, 1991. Unterberg, "National Self-determination," 638, observes an ironic contrast in the careers of George Washington and Abraham Lincoln. The fame of the first rested on his "successful conduct of a struggle for self-determination," the fame of the second "on his success in suppressing such a struggle."

Six: War and the Constitution:
Abraham Lincoln and Franklin D. Roosevelt—
Arthur M. Schlesinger, Jr.

1. Tyler Dennett, ed., *Lincoln and the Civil War in the Diaries and Letters of John Hay* (New York, 1939), 121.

2. Roy P. Basler, ed., Marion Dolores Pratt and Lloyd A. Dunlap, assoc. eds., *The Collected Works of Abraham Lincoln,* 9 vols. (New Brunswick, N.J., 1953–55), 3:29, 4:264.

3. *Franklin D. Roosevelt, Public Papers and Addresses . . . 1933* (New York, 1938), 14–15.

4. F. D. Wormuth and E. B. Firmage, *To Chain the Dog of War: The War Power of Congress in History and Law* (Dallas, 1984), 30.

5. Basler, ed., *The Collected Works of Lincoln,* 1:452.

6. C. C. Tansill, ed., *Documents Illustrative of the Formation of the Union of the American States* (Washington, 1927), 115.

7. Jefferson to W. C. Claiborne, February 3, 1807; to James Brown, October 27, 1808, Andrew A. Lipscomb, ed., Albert Ellery Bergh, managing ed., *The Writings of Thomas Jefferson,* 20 vols. (Washington, 1905), 11:151, 12:183.

8. Jefferson to J. B. Colvin, September 20, 1810, ibid., 12:418–22.

9. A. D. Sofaer, *War, Foreign Affairs and Constitutional Power* (Cambridge, Mass., 1976), 377–79.

10. Basler, ed., *Collected Works of Lincoln,* 4:426.

11. Adams in the House of Representatives, May 25, 1836, quoted in C. A. Berdahl, *War Powers of the Executive in the United States* (Urbana, 1921), 15.

12. Basler, ed., *Collected Works of Lincoln,* 6:408; 5:421, 6:29–30, 428.

13. Dennett, ed., *Hay Diaries,* 205.

14. Basler, ed., *Collected Works of Lincoln,* 7:281.

15. Ibid., 6:263, 267. The best work on the subject is Mark E. Neely, Jr., *The Fate of Liberty: Abraham Lincoln and Civil Liberties* (New York, 1991).

16. B. R. Curtis, "Executive Power" (1862), reprinted in B. R. Curtis, ed., *Memoir of Benjamin Robbins Curtis, with Writings* (Boston, 1879), vol. II.

17. Louis J. Jennings, *Eighty Years of Republican Government in the United States* (London, 1868), 36.

18. James Bryce, *The American Commonwealth,* 2 vols. (New York, 1888), 1:51,61.

19. Henry Adams, "The Sessions, 1869–1870," *North American Review,* 228 (July 1870) in Henry Adams, *The Great Secession Winter of 1860–61 and Other Essays,* George E. Hochfield, ed. (New York, 1958), 195.

20. Quoted by A. J. P. Taylor, *Essays in English History* (London, 1976), 13.

21. Basler, ed., *Collected Works of Lincoln,* 7:281.

22. 2 Black 635.

23. Churchill to Roosevelt, June 15, 1940, Warren F. Kimball, ed., *Churchill and Roosevelt: The Complete Correspondence* (Princeton, 1984), 1:51.

24. W. L. Langer and S. E. Gleason, *The Challenge to Isolation* (New York, 1952), 539.

25. Roosevelt to Churchill, May 16, 1940, Kimball, ed., *Churchill and Roosevelt . . . Correspondence,* 1:38.

26. R. H. Jackson, "Acquisition of Naval and Air Bases in Exchange for Over-age Destroyers," *Official Opinions of the Attorneys General of the United States* (Washington, 1941), 39:484–96; cf. also his concurring opinion in *Youngstown Co. v Sawyer* (1952).

27. E. S. Corwin, letter in *New York Times,* October 13, 1940.

28. *New York Times,* April 25, 1941.

29. See C. A. Beard, *American Foreign Policy in the Making 1932–1940* (New Haven, 1946) and *President Roosevelt and the Coming of the War 1941* (New Haven, 1948).

30. Robert A. Taft, *A Foreign Policy for Americans* (Garden City, N.Y., 1951), 31.

31. Henry L. Stimson and McGeorge Bundy, *On Active Service in Peace and War* (New York, 1948), 368.

32. Roosevelt to Grenville Clark, July 15, 1941, Roosevelt Papers, Franklin D. Roosevelt Library, Hyde Park, New York.

33. Roosevelt to R. H. Jackson, May 21, 1940; to T. H. Eliot, February 21, 1941, Roosevelt Papers.

34. Cabell Phillips, "No Witch Hunts," *New York Times Magazine,* September 21, 1941.

35. Francis Biddle, *In Brief Authority* (New York, 1962), 238.

36. Archibald Cox, *The Court and the Constitution* (Boston, 1987), 194–95.

37. *West Virginia Board of Education v. Barnette,* 319 U.S. 624.

38. American Civil Liberties Union, *Annual Report,* 1944.

39. Roosevelt, *Public Papers . . . 1942* (New York, 1950), 364.

40. Annual Message, January 6, 1941, Roosevelt, *Public Papers . . . 1940* (New York, 1941), 670.

41. *Ex parte Milligan,* 4 Wall. 2, 125 (1866).

42. Adams, "The Session, 1869–1870," 194–95.

43. Basler, ed., *Collected Works of Lincoln,* 8:152; Roosevelt, *Public Papers . . . 1942,* 365.

44. Basler, ed., *Collected Works of Lincoln,* 6:267.

45. *Korematsu v. U.S.,* 323 U.S. 14 (1944).

Seven: War Opponent and War President—
Gabor S. Boritt

1. Surely voicing the thoughts of many others who have lived at Gettysburg, the above notions were triggered in my mind by Robert Inman's novel *Home Fires Burning* (Boston, 1987), 208. In it Jake Tibbetts, the editor of a Southern weekly newspaper, writes: "No monument can honor warriors without honoring war itself."

2. A convenient place to read Einstein's letter is in Daniel Boorstin, ed., *An American Primer* (New York, 1968), 884–86. Richard Rhodes, *The Making of the Atomic Bomb* (New York, 1986), 314–15, 332, notes that no evidence indicates that Roosevelt ever read the Einstein letter though the President did learn of its contents. The architect and sculptor of the Vietnam memorial are, respectively, Maya Ying Lin and Frederick Hart. The Washington, Lincoln, and Einstein monuments were created, respectively, by Robert Mills, Daniel Chester French with Henry Bacon, and Robert Berks, though others, too, were involved.

3. Augustine's writing on war can be readily found in Albert Marrin, ed., *War and the Christian Conscience* (Chicago, 1971), 52–67; the quotation is from p. 55. This volume also provides a glimpse of the thought of St. Thomas Aquinas. For a recent attempt to summarize changing attitudes toward war, see Bernard Brodie, *War and Politics* (London, 1974), 223–75. For a focus on the ancients see Gerardo Zampaglione, *The Idea of Peace in Antiquity,* trs. Richard Dunn (Notre Dame, 1973).

4. Michael Howard, *War and the Liberal Conscience* (New Brunswick, 1986), 3. Cf. Michael Walzer, *Just and Unjust Wars: A Moral Argument with Historical Illustrations* (New York, 1977). Another fundamental assumption of this lecture is best expressed in the words of J. G. A. Pocock, provided that the word "just" is moderated in the following sentence: "What people claim to be doing and how they justify it is just as revealing as what they finally do." *Virtue, Commerce, and History* (Cambridge, 1985), 218.

5. John C. Nicolay and John Hay, *Abraham Lincoln, a History,* 10

vols. (New York, 1890), 1:27; Dennis Hanks to William H. Herndon, June 13, 1865, Herndon-Weik Papers, Library of Congress.

6. Parkman as quoted in Henry Dwight Sedgwick, *Francis Parkman* (Boston, 1904), 311.

7. Roy P. Basler, ed., Marion Dolores Pratt and Lloyd A. Dunlap, asst. eds., *The Collected Works of Abraham Lincoln,* 9 vols. (New Brunswick, 1953–55), 4:62. Something of a controversy exists about the import of the passage quoted above. Don E. Fehrenbacher, *Lincoln in Text and Context: Collected Essays* (Stanford, 1987), 224–26, questions the oedipal meanings given to the story in Charles B. Strozier, *Lincoln's Quest for Union: Public and Private Meanings* (New York, 1982), 25–26. See also *The Historian's Lincoln: Rebuttals. What the University Press Would Not Print* (Gettysburg, 1988), 19–23. The central fact, that Lincoln was not much of a hunter, remains.

8. Justin G. Turner and Linda Levitt Turner, *Mary Todd Lincoln: Her Life and Letters* (New York, 1972), 293, 296, 299; Basler, ed., *Collected Works,* 1:299–303. There is a substantial literature on both frontier violence and dueling.

9. Ibid., 4:64, 1:510.

10. The story appears untouched in W. D. Howells' *Life of Abraham Lincoln* (Bloomington, 1960), 38–39, which Lincoln corrected. See also W. H. Green to William H. Herndon, May 30, 1865 (cf. Reid W. A. Cury, n.d. Rec. group IV, reel 11, frame 2998), Herndon-Weik Papers; and Benjamin P. Thomas, *Lincoln's New Salem* (Springfield, 1934), 55.

11. Ibid., 56.

12. *Illinois House Journal,* 1840–41, 353; *Congressional Globe,* April 11, 1848, 30:1, 616. Lincoln's stance probably contained both anti-military and non-military elements. Of the varied works that touch on American attitudes toward matters military during this period, the most useful is Marcus Cunliffe, *Soldiers and Civilians: The Martial Spirit in America, 1775–1865* (Macmillan, 1973).

13. Basler, ed., *Collected Works of Lincoln,* 1:108–15, 3:316; and Chapter 1 *supra.* During the past generation a substantial discussion of the Lyceum address has taken place with contributions from Edmund

Wilson, Harry V. Jaffa, Glen E. Thurow, Laurence Berns, George B. Forgie, Dwight G. Anderson, Major L. Wilson, Marcus Cunliffe, Robert V. Bruce, Kenneth M. Stampp, and Richard N. Current. To avoid making the footnotes longer than the text, readers are asked to turn for a guide to the recent literature: Gabor S. Boritt, ed., Norman O. Forness, assoc. ed., *The Historian's Lincoln: Pseudohistory, Psychohistory, and History* (Urbana, 1988), and Boritt, ed., *The Historian's Lincoln: Rebuttals*.

14. Basler, ed., *Collected Works of Lincoln*, 1:112, 279, 115, 114.

15. Ibid., 4:235, 240.

16. Ibid., 1:114.

17. Ibid., 1:278. For American views of the Revolution, including those in Lincoln's time, see Michael Kammen, *A Season of Youth: The American Revolution and the Historical Imagination* (New York, 1978).

18. Ibid., 1:439, 451–52, 447.

19. Quoted in Usher F. Linder, *Reminiscences of the Early Bench and Bar in Illinois* (Chicago, 1879), 87. See also Joseph Gillespie's postscript, ibid., 404–5; Tyler Dennett, ed., *Lincoln and the Civil War in the Diaries and Letters of John Hay* (New York, 1939), 80; *Illinois Register* (Springfield), Oct. 1, 1847, June 21, 1849.

20. The quotations above are identified and the literature on Lincoln's opposition to the Mexican War is examined in G. S. Boritt, "A Question of Political Suicide? Lincoln's Opposition to the Mexican War," *Journal of the Illinois State Historical Society*, 57 (1974): 79–100. See also Mark E. Neely, Jr., "Lincoln and the Mexican War: An Argument by Analogy," *Civil War History*, 24 (1978):5–24. Mid-nineteenth-century American attitudes toward this war are examined in Robert W. Johannsen, *To the Halls of the Montezumas: The Mexican War in the American Imagination* (New York, 1985).

21. Basler, ed., *Collected Works of Lincoln*, 3:471–72. See also Charles De Benedetti, *The Peace Reform in American History* (Bloomington, 1980). For the literature of peace studies see Charles F. Howlett and Glen Zeitzer, *The American Peace Movement: History and Historiography* (Washington, 1985).

22. Basler, ed., *Collected Works of Lincoln*, 4:64.

23. Ibid., 1:508.

24. The quotations are from ibid., 1:507, 477, 2:137.

25. Ibid., 1:439.

26. Ibid., 1:509–10.

27. Ibid., 2:149.

28. David R. Locke in Allen Thorndike Rice, ed., *Reminiscences of Abraham Lincoln by Distinguished Men of His Time* (New York, 1888), 442.

29. Basler, ed., *Collected Works of Lincoln,* 2:149–50.

30. Ibid., 149.

31. Ibid., 3:318–19. It is possible to interpret the materials above differently. If students have not already done so they surely will. For example, Lincoln's recognition of the horrors of the American Revolution can be tied to his jealousy toward George Washington's fame. Lincoln's denunciation of the Mexican War can be seen as little more than playing politics. Opposition to Democratic generals while supporting Whig ones can be characterized as hypocrisy. Making fun of matters military might be little more than making fun, or perchance making political hay. Ultimately, then, a historian's interpretation of Lincoln, or any subject, depends to a fair degree on his/her overall judgment and understanding of that subject.

32. Ibid., 1:205, 2:282, 6:410.

33. Ibid., 1:184, 209–10; Mark E. Neely, Jr., "Some Curiosities of a Congressional Career," *Lincoln Lore,* 1688 (1977).

34. For a discussion of the genre of eulogies, as well as Lincoln in particular, see Mark E. Neely, Jr., "American Nationalism in the Image of Henry Clay: Abraham Lincoln's Eulogy of Henry Clay in Context," *Register of the Kentucky Historical Society,* 73 (1975):31–60.

35. Basler, ed., *Collected Works of Lincoln,* 2:127.

36. Ibid., 2:86, 85, 89.

37. Ibid., 85; 1:514.

38. Ibid., 279.

39. For revolutions, see Thomas J. Pressly, "Bullets and Ballots: Lincoln and the 'Right of Revolution.' " *American Historical Review,* 67 (1962):647–62.

40. Basler, ed., *Collected Works of Lincoln,* 1:109.

41. Ibid., 1:386–89; 8:332.

42. Robert V. Bruce, *Lincoln and the Tools of War* (Indianapolis, 1956).

43. T. Harry Williams, *Lincoln and His Generals* (New York, 1952), viii.

44. Basler, ed., *Collected Works of Lincoln,* 5:426; 4:259; Williams O. Stoddard, *Inside the White House in War Time* (New York, 1890), 178–79.

45. Chapter 2, supra; cf. Charles B. Strozier, *Unconditional Surrender and the Rhetoric of Total War: From Truman to Lincoln* (New York, 1987); Mark E. Neely, Jr., "Was the Civil War a Total War?," *Civil War History,* 37 (1991):5–28; and Herman Hattaway and Archer Jones, "Lincoln as a Military Strategist," ibid., 26 (1980):293–303.

46. Chapter 3, supra.

47. Basler, ed., *Collected Works of Lincoln,* 4:439.

48. Noah Brooks, "A Boy in the White House," *St. Nicholas,* 10 (Nov. 1882); 59; and *Washington D.C. in Lincoln's Time,* Herbert Mitgang, ed. (Chicago, 1971), 196.

49. Michael Shaara, *The Killer Angels. A Novel* (New York, 1974).

50. Marquis Adolphe de Chambrun, *Impressions of Lincoln and the Civil War: A Foreigner's Account* (New York, 1952), 84. Cf. Basler, ed., *Collected Works of Lincoln,* 4:271.

51. John Henry Cramer, *Lincoln Under Enemy Fire: The Complete Account of His Experiences During Early's Attack on Washington* (Baton Rouge, 1948); *Diaries of John Hay,* 208, 209. A brief review of the literature that has grown up around the incident appears in Liva Baker, *The Justice from Beacon Hill: The Life and Times of Oliver Wendell Holmes* (New York, 1991), 151–52. See also Sheldon M. Novick, *Honorable Justice: The Life of Oliver Wendell Holmes* (Boston, 1989), 422n.37.

52. John Minor Botts, *The Great Rebellion* (New York, 1866), 196. See also Basler, ed., *Collected Works of Lincoln,* 4:237, 243; 5:165; 8:1.

53. Ibid., 6:500.

54. Ibid., 7:394; 8:1, 7:395.

55. Ibid., 8:333.

56. J. Rufus Fears, ed., *Selected Writings of Lord Acton*, 3 vols. (Indianapolis: Liberty Classics, 1985–88), 1:50. The present lecture provides only a bare outline of Lincoln's views on war. I hope that all its parts as well as the whole will be re-examined and challenged by future scholars.

57. For arguments against the bomb, see John Finnis, Joseph M. Boyle, Jr., and Germain Grisez, *Nuclear Deterrence, Morality and Realism* (New York, 1987), and Oliver O'Donovan, *Peace and Certainty: A Theological Essay on Deterrence* (Grand Rapids, 1988).

58. Richard N. Current, *The Lincoln Nobody Knows* (New York, 1958), 186; Basler, ed., *Collected Works of Lincoln*, 5:53.

Contributors

Gabor S. Boritt, Robert C. Fluhrer Professor of Civil War Studies and Director of the Civil War Institute, Gettysburg College, is the author of *Lincoln and the Economics of the American Dream* (1978). His most recent book is *Why the Confederacy Lost* (1992).

Robert V. Bruce, Professor of History, Boston University, won the Pulitzer Prize for his most recent book, *The Launching of Modern American Science, 1846–1876* (1987).

David Brion Davis, Sterling Professor at Yale University, former president of the Organization of American Historians, won the Pulitzer Prize for *The Problem of Slavery in Western Culture* (1966). His most recent book is *Revolutions: Reflections on American Equality and Foreign Liberations* (1990).

Carl N. Degler, Margaret Byrne Professor of American History at Stanford University, former president of the American Historical Association, the Organization of American Historians, and the Southern Historical Association, won the Pulitzer Prize for *Neither Black Nor White* (1971). His most recent book is *The Search for Human Nature: The Decline and Revival of Darwinism in American Social Thought* (1991).

James M. McPherson, George Henry Davis '86 Professor of American History, Princeton University, won the Pulitzer Prize for *Battle Cry of Freedom* (1988). His most recent book is *Ordeal by Fire: The Civil War and Reconstruction* (2nd ed., 1992).

Arthur M. Schlesinger, Jr., Albert Schweitzer Professor of Humanities, City University of New York, won the Pulitzer Prize for both *The Age of Jackson* (1945) and *The Age of Roosevelt* (1957). His most recent book is *The Disuniting America* (1992).

Kenneth M. Stampp, Morrison Professor of American History, Emeritus, University of California, Berkeley, is former president of the Organization of American Historians. His most influential book is *The Peculiar Institution* (1956). His most recent one, is *America in 1857: A Nation on the Brink* (1990).